D1565874

NONLINEAR DYNAMICAL SYSTEMS

To my students, in the hope that they may
learn from me as much as I learn from them

Felix qui potuit rerum
cognoscere causas

Virgil

Nonlinear Dynamical Systems

P.A. COOK

Control Systems Centre
University of Manchester Institute of Science and Technology
Manchester, England

ENGLEWOOD CLIFFS, NEW JERSEY LONDON MEXICO NEW DELHI
RIO DE JANEIRO SINGAPORE SYDNEY TOKYO TORONTO WELLINGTON

Library of Congress Cataloging in Publication Data

Cook, P. A. (Peter A.)
Nonlinear dynamical systems.

Bibliography: p.
Includes index.
1. System analysis. 2. Control theory.
3. Nonlinear theories. I. Title.
QA402.C595 1986 629.8′312 85–19283
ISBN 0–13–623216–7

British Library Cataloguing in Publication Data

Cook, P. A. (Peter A.)
Nonlinear dynamical systems.——(Prentice-Hall International series in control engineering)
1. Control theory 2. Nonlinear theories I. Title
629.8′36 QA402.3

ISBN 0–13–623216–7

Prentice-Hall, Inc., *Englewood Cliffs, New Jersey*
Prentice-Hall International (UK) Ltd, *London*
Prentice-Hall of Australia Pty Ltd, *Sydney*
Prentice-Hall Canada, Inc., *Toronto*
Prentice-Hall of India Private Ltd, *New Delhi*
Prentice-Hall of Japan, Inc., *Tokyo*
Prentice-Hall of Southeast Asia Pte Ltd, *Singapore*
Editora Prentice-Hall do Brasil Ltda, *Rio de Janeiro*
Whitehall Books Ltd, *Wellington, New Zealand*

Phototypeset by Gecko Limited, Bicester, Oxon
Printed in Great Britain by
A. Wheaton & Co. Ltd, Exeter
1 2 3 4 5 90 89 88 87 86

ISBN 0-13-623216-7

CONTENTS

PREFACE

This work is largely, though not entirely, based on lectures which I have given, over several years, to postgraduate students in the Control Systems Centre at the University of Manchester Institute of Science and Technology. It is intended as a textbook to be used in M.Sc. or final-year B.Sc. courses, in control or electrical engineering or mathematics, but also as an introduction to more advanced topics, suitable for students wishing to do research in this area. With the teaching aspect in mind, I have included a number of worked examples in the text, and also exercises for the reader, at the ends of most of the chapters, with solutions provided. For the benefit of prospective research workers, there is an appendix containing notes on the origin of the material in the chapters, together with a list of references to the technical literature, and all such references have in fact been placed in the appendix, not in the main text.

The motivation for studying nonlinear dynamical systems lies partly in their practical importance and partly in the fascination of their richly varied behaviour. From the former viewpoint, the development has been principally aimed at applications in control engineering, involving the dynamical properties of nonlinear feedback systems, although some of the examples are taken from other fields where similar problems also arise. At the same time, I have tried to give an indication of what is known, and what remains to be discovered, on the theoretical side of the subject, within the limitations imposed by my desire to keep the mathematics at as elementary a level as possible. Nevertheless, the main objective has been to present methods which can be used to gain an understanding of how nonlinear systems behave, rather than a fully rigorous treatment, for which the interested reader is advised to consult the relevant literature.

Although this is far from being the first book of its kind, I hope that its appearance may be justified by the continually developing state of the subject, if by nothing else. For stimulating and maintaining my interest in this field, and in related areas of control and system science, I am greatly indebted to my colleagues in the Control Systems Centre. Finally, for the rapid and skilful production of the typescript, it is a pleasure to thank Mrs Joy Munro.

P.A.C.

1 INTRODUCTION

It is not easy to give a satisfactory general definition of a dynamical system, nor will it be necessary for our purposes. Nevertheless, since we have to begin somewhere, let us take the following loose description. A dynamical system is characterised by a set of related variables, which can change with time in a manner which is, at least in principle, predictable provided that the external influences acting on the system are known. In this book, we shall be concerned only with deterministic systems, that is to say, we take no account of statistical properties. Further, we shall assume, in any particular case, that we are provided with an accurate model of the system in question and do not need to concern ourselves about how it was obtained. On the other hand, since we know that, in practice, all models are approximate and all systems are subject to stochastic effects, we shall need to keep in mind the question of how reliable our methods and results can be expected to be, in the face of these uncertainties.

The term 'dynamics' evokes those branches of physics in which it has become a familiar concept, for example, classical mechanics and electromagnetic theory, but there is no reason why we should restrict our attention to these areas. One can just as well study the dynamics of biological, social or economic systems, provided that one can obtain adequate models to describe their behaviour in response to the forces which drive them. This, however, is a large proviso, and the relative lack of quantitative development in these fields means that the majority of applications which concern us are to be found in the physical sciences and engineering. From an engineering viewpoint, a prime reason for wishing to understand the behaviour of a system is to enable us to control it, and conversely, if we wish to exercise control, we must have some idea how the system will respond to whatever influence we can exert upon it. It is thus natural that the study of system behaviour should go hand in hand with that of control, to the extent that the whole area (sometimes known as cybernetics) is nowadays often referred to under the title 'systems and control'.

In order to make any quantitative progress in understanding a system, we require a mathematical model. Such models may, as we shall see, be formulated in many ways, but their essential feature is to enable us to predict the system's future behaviour, given its initial condition and a knowledge of the external forces which affect it. The mathematical structure most naturally adapted to this purpose is the so-called state-space representation, which consists of a set of differential equations, describing the evolution of the variables whose values at any given instant determine the current state of the system. These are known as the state variables and their values at any particular time are supposed to contain sufficient information for the future evolution of the system to be predicted, given that the external influences (or input variables) which act upon it are known. The differential equations must therefore be of first order in the time-derivative, so that the initial values of the variables will suffice to determine the solution. For convenience of notation, it is usual to collect the state variables into a vector $\mathbf{x}$ (the state vector), the input variables into a vector $\mathbf{u}$ (the input vector), and write the equations in the form

$$\dot{\mathbf{x}} = \mathbf{f}(\mathbf{x},\mathbf{u},t)$$

where the dot denotes differentiation with respect to time (t) and the function $\mathbf{f}$ is in general nonlinear. These equations are to be solved for $\mathbf{x}$ as a function of t, given the value of $\mathbf{x}$ at some initial instant $t = t_0$ and the subsequent values of $\mathbf{u}$ as a function of t, for $t \geqslant t_0$. To be sure that this is a mathematically valid representation, we should place certain restrictions on the nonlinear function $\mathbf{f}$, in order to guarantee the existence and uniqueness of solutions. We shall sometimes find it convenient, however, to use idealised models in which the functions have certain features (for example, discontinuities) which can cause difficulties in this respect, and then it will be necessary to examine the analysis carefully, so as to ensure that such problems do not invalidate the solutions. In the control engineering context, it is also appropriate to distinguish another set of quantities (called output variables) which represent aspects of the system's behaviour that can be measured and controlled. These are usually a subset of the state variables but may in general depend on $\mathbf{u}$ and t as well, so we collect them into another vector $\mathbf{y}$ (the output vector) and write

$$\mathbf{y} = \mathbf{h}(\mathbf{x},\mathbf{u},t)$$

where $\mathbf{h}$ is another function which may be nonlinear. By solving the differential equations for $\mathbf{x}(t)$, we thus obtain the dependence of $\mathbf{y}(t)$ on $\mathbf{x}(t_0)$ and on $\mathbf{u}(\tau)$ for $t_0 \leqslant \tau \leqslant t$.

The basic state-space description formulated above may be generalised in a number of ways. For example, systems described by partial-differential equations (distributed-parameter systems) can be represented in state-space form by allowing $\mathbf{x}$ (and possibly $\mathbf{u}$ and $\mathbf{y}$ also) to contain an infinite number of components. Such infinite-dimensional systems, however, present various subtle mathematical difficulties and are therefore usually approximated in practice by finite-dimensional (lumped-parameter) models of the kind we have just considered. Another, more widely used, modification is the introduction of delay effects, which may arise either as a result of physical phenomena, such as material transport, or in an attempt to represent complicated dynamical features in a simple way. This results in the differential equations for the state vector being replaced by delay-differential (or hysterodifferential) equations which, in the simplest case (a single delay) take the form

$$\dot{\mathbf{x}}(t) = \mathbf{f}(\mathbf{x}(t),\mathbf{x}(t-T),\mathbf{u}(t),\mathbf{u}(t-T),t)$$

where T (the delay) is a positive constant. More generally, $\mathbf{f}$ may contain further arguments corresponding to delays of different lengths. Despite their superficial resemblance to the state-space form, however, such equations do not strictly belong to this category, since the solution is not uniquely determined by the initial value $\mathbf{x}(t_0)$, but rather by the function $\mathbf{x}(\tau)$ over the interval $t_0-T \leqslant \tau \leqslant t_0$, where T is the maximum delay associated with $\mathbf{x}$. Up to now, we have considered only continuous-time systems, that is to say, those in which t is regarded as a continuous variable, but it is sometimes useful, for instance in the analysis of sampled-data control systems, to restrict attention to the values of variables at a discrete sequence of instants, such as when t is an integer. This generates a discrete-time system, whose state-space representation is a set of difference equations (or recurrence relations) of the form

$$\mathbf{x}(t+1) = \mathbf{f}(\mathbf{x}(t),\mathbf{u}(t),t)$$

to be solved recursively for $\mathbf{x}$.

As an alternative to the state-space form, or in conjunction with it, we shall often use the input–output description, which relates the input and output variables of a system directly. For example, a system with a scalar input and output might be described by a single nth order differential equation involving only $\mathbf{u}$ and $\mathbf{y}$, which could be obtained by eliminating $\mathbf{x}$ from the state–space description, consisting of n first-order differential equations and one algebraic equation giving $\mathbf{y}$ as a function of $\mathbf{x}$ and $\mathbf{u}$. Conversely, starting from the input-output relation, a state-space representation could be constructed by introducing a suitable set of n auxiliary quantities to serve as state variables, and the ability to pass readily from one description to the other will turn out to be very useful. In the case of static (nondynamical) relations, state variables are irrelevant and the input–output form is natural to represent, for instance, a pure time delay

$$\mathbf{y}(t) = \mathbf{u}(t-T)$$

or an instantaneous functional relationship

$$\mathbf{y} = \mathbf{h}(\mathbf{u})$$

where $\mathbf{h}$ is a general nonlinear function (or, even more generally, a multivalued relation). A convenient feature of the input–output viewpoint is that it lends itself naturally to a diagrammatic representation, in which a system is shown as a box with arrows entering (to denote input variables) and leaving it (output variables). By combining boxes corresponding to different subsystems, we can thus obtain the block diagram representation of a complex system, familiar to all students of control engineering. If the input–output relation of a subsystem is linear and constant (time-invariant), the corresponding block can be labelled with its transfer function, obtained by the use of Laplace transforms. On the other hand, a static nonlinear function can be represented by a block showing the graph (or characteristic) of the relation between its input and output variables, and there are also special symbols to denote summation and multiplication, as in Fig. 1.1.

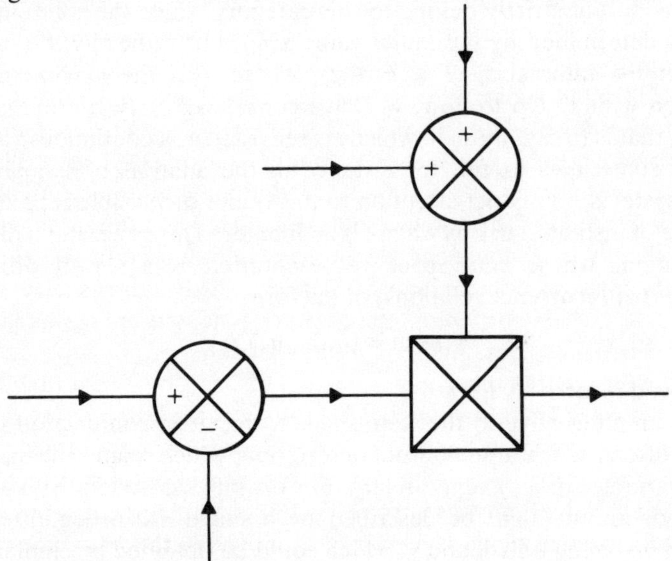

Fig. 1.1 Representation of sums, differences and products.

1.1 Types of nonlinearity

Nonlinear functions may arise in a dynamical model either because they are intrinsic to the nature of the system or because, in a technological case such as a control system, they have been deliberately introduced by the designer for a specific purpose. The variety of possible nonlinearities is infinite, but it may nevertheless be worthwhile to classify them into some general categories, with features which permit (or preclude) the application of particular analytical methods.

In the first place, we may have simple analytic functions such as powers, sinusoids and exponentials of a single variable, or products of different variables. A significant feature of these functions is that they are smooth enough to possess convergent Taylor expansions at all points and consequently can be linearised, that is to say, approximated by linear expressions in the neighbourhood of any given operating point, so that the full mathematical power of linear system analysis becomes available. Besides arising naturally in phenomenological modelling, nonlinearities of this type may also appear as a deliberate feature of system design. For example, in adaptive control systems, where the control parameters, instead of being fixed, are made to vary in accordance with the performance of the system, product terms occur automatically in the equations. There is also an important class of systems called bilinear systems, in which the only nonlinear terms consist of state variables multiplied by input variables, and these have been widely studied since they constitute, in a sense, the simplest possible generalisation of linear systems, being linear in the state and input separately. If a system model contains only analytic nonlinearities, it also admits the possibility of using a special type of input–output representation, in which the output is expressed as a kind of generalised power series (Volterra series) containing multiple convolution integrals of products of the input variables evaluated at different times.

Another type of nonlinear function frequently used in system modelling is the piecewise-linear approximation, which consists of a set of linear relations valid in different regions. Such functions are not analytic at all points, since they contain discontinuities of value or gradient, but they have the advantage that the dynamical equations become linear (and hence soluble) in any particular region, and the solutions for different regions can then be joined together at the boundaries. Approximations of this type are widely used in control engineering to model the behaviour of actuating devices such as valves, motors and relays. For example, the response of an actuator to a control signal is always limited by saturation, which can be modelled by a piecewise-linear function as in Fig. 1.2. Similarly, many actuators will not respond at all until the input signal reaches a certain magnitude, so that the input–output characteristic contains a dead zone, which can be represented by a piecewise-linear relation as in Fig. 1.3. The functions involved in these examples are continuous in value, although not in slope, but discontinuous functions are also employed. The simplest case is the so-called ideal relay characteristic, which is mathematically represented by the signum function

$$y = \operatorname{sgn}(u)$$

shown in Fig. 1.4. Besides representing the behaviour of a relay or

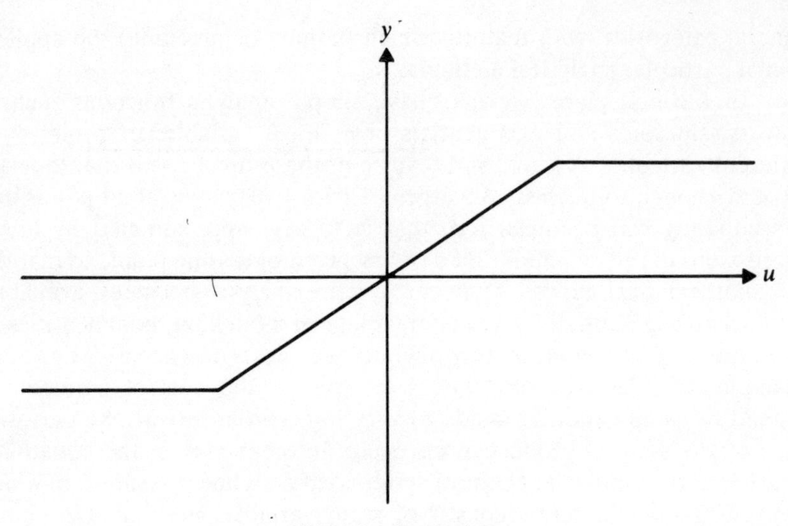

Fig. 1.2 Saturation.

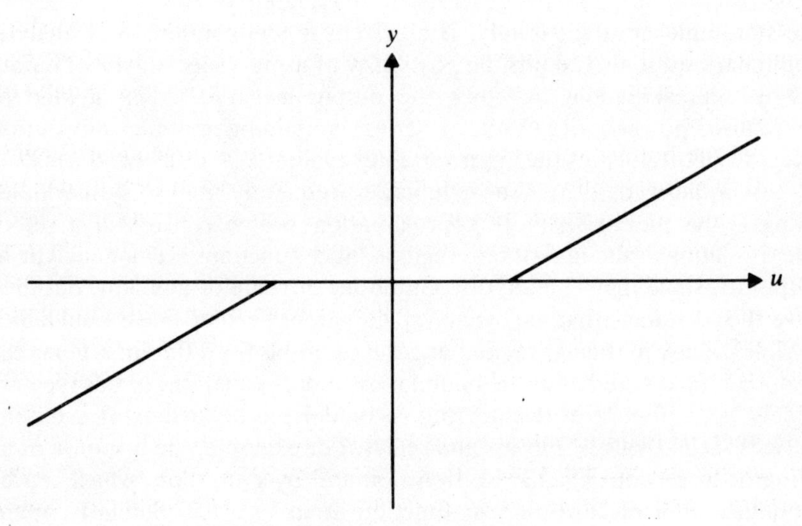

Fig. 1.3 Dead zone.

switching device, it can also be used to approximate the effect of Coulomb friction. As already mentioned, we need to be careful when using discontinuous approximants, but they arise quite naturally in many applications. These include some modern types of control scheme, such as 'variable-structure' systems, where the control parameters are switched when the state variables reach certain values, and 'voting' strategies (highest wins or lowest wins) where the parameter values are determined by which of several error signals has the greatest or least magnitude. Moreover, when the behaviour of this kind of system is analysed in detail,

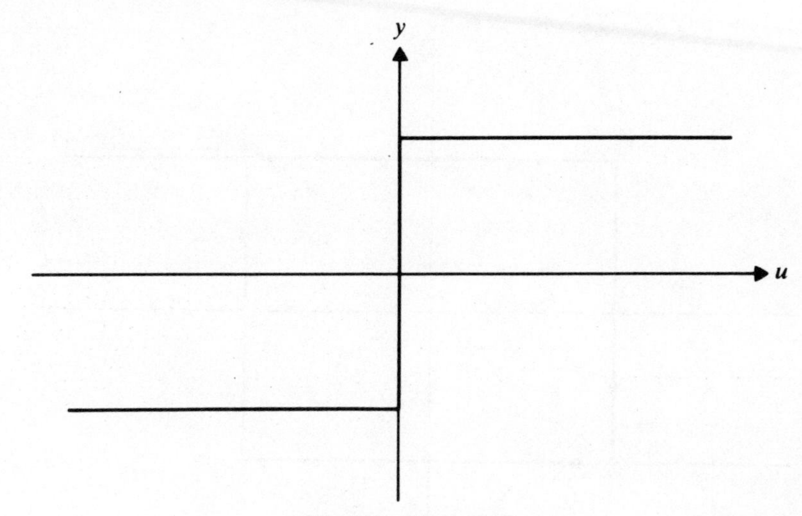

Fig. 1.4 Ideal relay.

it turns out that the mathematical ambiguities encountered at the switching boundaries can indeed correspond to real physical effects, manifested in practice as rapid small oscillations (or chatter) across the points of discontinuity.

All the nonlinearities considered so far have been genuine (single-valued) functions, but we shall also have occasion to make use of what are loosely called 'multivalued' functions or, more accurately, relations. These are used principally for the modelling of hysteresis, either in the physical response of a magnetic or elastic material, or to represent the non-ideal behaviour of a practical relay device which switches off at a different point from the one where it switches on, giving rise to an overlap region in its characteristic, as shown in Fig. 1.5. A relation of this type thus consists of two or more branches, together with a prescription for switching from one to another, so that the output value depends on the history of the input as well as on its current value. It is therefore sometimes said to have 'memory', and a single-valued nonlinearity is correspondingly called 'memoryless'. Another example of a multivalued nonlinearity is the 'backlash' characteristic used in modelling the behaviour of gear trains, which in its simplest form (friction controlled) is illustrated in Fig. 1.6. This relation is in effect infinitely-many-valued since, for a given drive position (input), the load position (output) can take any value between two extremes. From the mathematical point of view, multivaluedness does not necessarily raise any essentially new difficulty, provided that the switching conditions are unambiguous. It does, however, make the state-space formulation more complicated, since the system now has to be represented by a combination of several (possibly infinitely many) different state-spaces, corresponding to the various branches of the nonlinearities. For

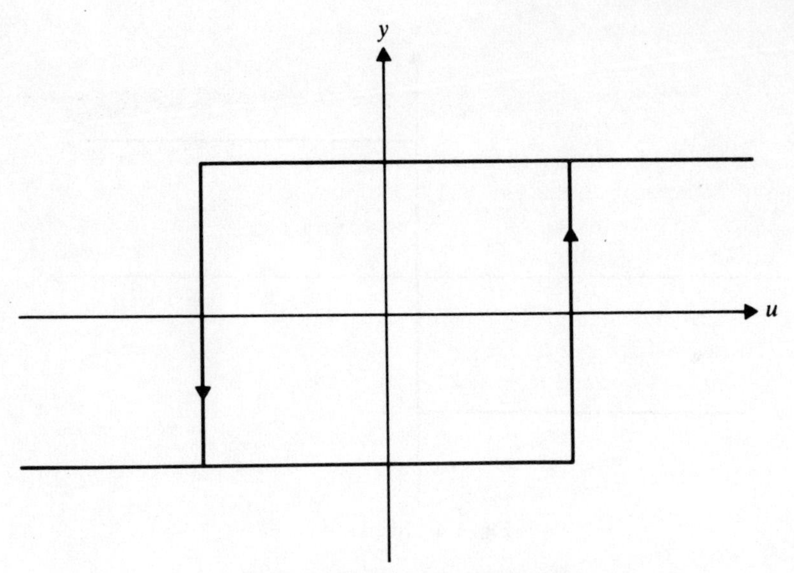

Fig. 1.5 Relay with hysteresis.

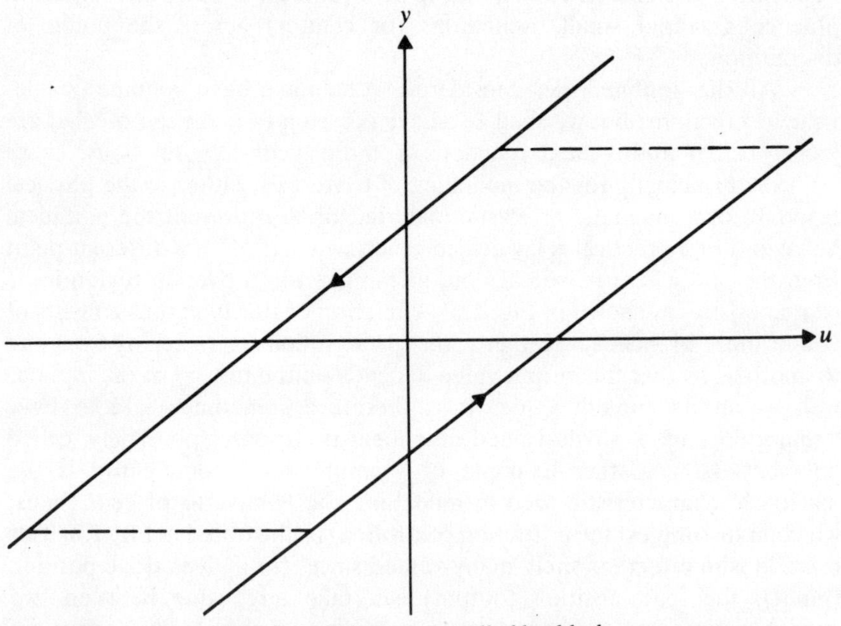

Fig. 1.6 Friction-controlled backlash.

infinitely-valued relations, as in the case of backlash or related phenomena such as stiction (stick-slip), it is in fact preferable to use an alternative description, involving state-spaces of different dimensions corresponding, for instance, to sticking and slipping modes, respectively.

1.2 Linear systems

Although we are principally concerned with nonlinear phenomena, it is appropriate at this point to review the special case of linear systems, partly because the mathematical analysis is then so much more tractable, and also because linear approximations are so widely applicable, to the extent that, for example, virtually all control system design methods are based upon them. Indeed, it is arguable that all effective approaches to the study of nonlinear systems make use, in some way, of approximately linear relationships and analogies with linear behaviour.

If we restrict our attention, for simplicity, to systems with finite-dimensional state-space representations, the equations describing a linear model become

$$\dot{\mathbf{x}} = \mathbf{A}\mathbf{x} + \mathbf{B}\mathbf{u}$$
$$\mathbf{y} = \mathbf{C}\mathbf{x} + \mathbf{D}\mathbf{u}$$

where $\mathbf{A}$, $\mathbf{B}$, $\mathbf{C}$ and $\mathbf{D}$ are matrices (possibly time-dependent) of appropriate dimensions. The great advantage of linearity is that, even in the time-dependent case, a formal solution can immediately be constructed, which is moreover applicable for all initial conditions and all input functions. For this purpose, we define the state transition matrix $\Phi(t,t_0)$ as the solution of the matrix differential equation

$$\Phi(t,t_0) = \mathbf{A}(t)\Phi(t,t_0)$$

with initial condition

$$\Phi(t,t_0) = \mathbf{I}$$

which exists and is unique for all matrix functions $\mathbf{A}(t)$ of practical interest (continuity being more than sufficient). The solution for $\mathbf{x}(t)$ is then given by

$$\mathbf{x}(t) = \Phi(t,t_0)\,\mathbf{x}(t_0) + \int_{t_0}^{t} \Phi(t,\tau)\,\mathbf{B}(\tau)\,\mathbf{u}(\tau)\,d\tau$$

and $\mathbf{y}(t)$ can be obtained immediately by substitution. This result, sometimes known as the 'variation of constants' formula, clearly exhibits the linear dependence of the solution on both the initial state and the input vector. The difficulty of solving the equation for the state transition matrix in general, however, prevents us from making much further progress with time–dependent systems, except in special cases as, for instance, when $\mathbf{A}(t)$ is periodic.

For time-invariant systems, on the other hand, that is to say, when $\mathbf{A}$, $\mathbf{B}$, $\mathbf{C}$ and $\mathbf{D}$ are constant matrices, a great simplification arises, because the state transition matrix depends essentially on only one variable rather than two. We can, in fact, without loss of generality, take the initial instant as the time origin ($t_0 = 0$), and we then have

$$\Phi(t,0) = \exp(\mathbf{A}t)$$

where the matrix exponential function is defined by the same formal power series expansion as for a scalar exponential, so that

$$\exp(\mathbf{M}) = \mathbf{I} + \sum_{k=1}^{\infty} \left(\frac{\mathbf{M}^k}{k!} \right)$$

which converges for any square matrix $\mathbf{M}$. This enables the solution of the state-space equations to be written as

$$\mathbf{y}(t) = \mathbf{C} \exp(\mathbf{A}t)\, \mathbf{x}(0) + \int_{t_0}^{t} \mathbf{g}(t-\tau)\, \mathbf{u}(\tau)\, \mathrm{d}\tau$$

where $\mathbf{g}(t)$ is the impulse response matrix, given by

$$\mathbf{g}(t) = \mathbf{C} \exp(\mathbf{A}t)\mathbf{B} + \mathbf{D}\delta(t)$$

with $\delta(t)$ denoting the Dirac delta-function (distribution) although in most cases this term is absent since $\mathbf{D} = 0$ (no instantaneous response). The above solution, however, is still somewhat awkward to use directly on account of the convolution integral involved, and it is more conveniently expressed with the aid of the Laplace transformation. Denoting the Laplace transforms of $\mathbf{u},\mathbf{y},\mathbf{g}$ by $\mathbf{U},\mathbf{Y},\mathbf{G}$, respectively, we obtain

$$\mathbf{Y}(s) = \mathbf{C}(s\mathbf{I}-\mathbf{A})^{-1}\, \mathbf{x}(0) + \mathbf{G}(s)\, \mathbf{U}(s)$$

where the transfer function matrix $\mathbf{G}(s)$ is given by

$$\begin{aligned} \mathbf{G}(s) &= \int_0^{\infty} \mathbf{g}(t) \exp(-st)\, \mathrm{d}t \\ &= \mathbf{C}(s\mathbf{I}-\mathbf{A})^{-1}\mathbf{B} + \mathbf{D} \end{aligned}$$

which reduces the solution of the equations to algebra instead of analysis. It also has the advantage of being directly related to the frequency-response of the system, that is to say, its response to a sinusoidal input signal, which is obtained from $\mathbf{G}(s)$ evaluated on the imaginary axis in the s-plane. Moreover, it generalises fairly readily to cases in which there is no finite-dimensional state-space, such as time–delay and distributed-parameter systems, where $\mathbf{G}(s)$ becomes non-rational, with a delay of length T, for example, being represented by a factor $\exp(-sT)$.

One property of dynamical systems, which is of crucial importance and which takes a particularly simple form in the linear case, is that of stability. Considering first the free or unforced case, that is to say, with no input applied to the system so that its response arises only from the initial conditions, we regard it as stable if all dynamical variables remain bounded as $t \to \infty$, and asymptotically stable if they all converge to zero in this limit. For a finite-dimensional time–invariant system, the unforced response is given by

$$\mathbf{x}(t) = \exp(\mathbf{A}t)\, \mathbf{x}(0)$$

whence the necessary and sufficient condition for asymptotic stability is that

$$\exp(\mathbf{A}t) \rightarrow 0$$

as $t \rightarrow \infty$, which is equivalent to the statement that every eignevalue of the 'plant' matrix **A** has negative real part. In the time-dependent case, the corresponding condition is

$$\Phi(t, t_0) \rightarrow 0$$

as $t \rightarrow \infty$, for all t_0, but the connection with eigenvalues of the plant matrix no longer holds. The most significant point about these conditions, however, is that they depend only on the system itself and not on its initial state. This is a specific consequence of linearity and remains valid for all types of linear system though not, in general, for nonlinear systems. Moreover, essentially the same conditions hold for stability in the input–output sense, which is applicable to forced systems. In this regard, a system is said to be stable if a bounded input always produces a bounded output, which is true, in the finite-dimensional time–invariant case, if and only if

$$\mathbf{g}(t) \rightarrow 0$$

as $t \rightarrow \infty$. Expressed in terms of Laplace transforms, this means that all the poles of $\mathbf{G}(s)$ must lie in the open left half-plane (Re $s < 0$), which relates directly to the asymptotic stability condition, since poles of the transfer function matrix can only occur at the roots of

$$\det(s\mathbf{I} - \mathbf{A}) = 0$$

which are just the eigenvalues of the plant matrix **A**.

1.3 Aspects of nonlinear behaviour

Because linear systems are so much easier to handle mathematically, the first step in dealing with a nonlinear system is usually, if possible, to linearise it around some nominal operating point. Assuming that the deviations from this condition are not too large in practice, the linearised approximation may well be adequate as a basis for analysis and design over a limited range of operation. Control systems are, in fact, normally designed initially in this way, and the effects of departures from linearity are then investigated by simulation methods, the result usually being that the operation of the controller tends to make the system more, rather than less, nearly linear in its behaviour. On the other hand, if the system is required to operate under a wide range of conditions, the fact that it is nonlinear will inevitably reveal itself through the dependence of the linearised model's parameters on the operating point about which it has been obtained. In order to apply linear design techniques, it may then be necessary to obtain a set of approximate models, by linearising around different operating points, and hence generate a sequence of controllers, to be brought into operation successively as the system passes through conditions where the corresponding models are approximately valid (scheduled control).

Another point which must be kept in mind, when considering large deviations from the nominal operating condition, is that the stability properties of a nonlinear system are essentially more complicated than in the linear case, and in particular, we have to distinguish between local and global aspects. For a linear system there is no such distinction, but when nonlinearities are present, several new features can appear. To begin with, the stability of a nonlinear system in the neighbourhood of an equilibrium point does not necessarily imply any global property. There may indeed be many equilibria, some stable and others not, in which case there will be only a limited region of convergence (domain of attraction) around any equilibrium point which is locally asymptotically stable. Furthermore, there can be other, essentially nonlinear, modes of behaviour, such as the persistent oscillations known as limit cycles, which constitute a kind of dynamic, rather than static, equilibrium and have no true counterpart in linear systems. In fact, there are much more complicated possibilities than the periodic behaviour associated with simple limit cycles, even to the extent of a quasi-stochastic situation arising, despite the deterministic nature of the system equations. This phenomenon, known as chaos, is particularly prevalent in discrete-time systems and can occur even for quite simple low-order models. In any case, the type of behaviour actually manifested by a nonlinear system, whether stable, unstable, oscillatory or chaotic, may depend critically on the input applied to it, in contrast to the linear case where all the dynamical properties can be described, for example by a transfer function, independently of the input.

The nonlinear nature of a system's input–output relation may show itself through performance degradation, limit cycle excitation or loss of stability for large input signals. It also introduces new features into the frequency-response properties, which are of great importance because of the central role which they play in the study of linear systems. In the linear case, the application of a sinusoidal input can only generate a sinusoidal output of the same frequency, amplified by a gain factor and phase-shifted through an angle, both of which can be obtained directly from the transfer function. With a nonlinear system, however, even if the output has the same period as the input, it will in general contain higher harmonics of the fundamental frequency, instead of being a pure sine-wave. More generally, it can also contain components at fractions of the input frequency (subharmonics) and even at quite unrelated frequencies, associated with internal dynamical phenomena such as limit cycles. Nevertheless, the frequency-response is still a useful tool, mainly because the harmonic content of the output signal is usually dominated by a small number of frequencies. In fact, it is often sufficient to consider only the fundamental frequency, in which case the input–output relation is represented by a nonlinear analogue of the transfer function, called a 'describing function', which depends on the input amplitude. Although this approximation must be used with care, as it can give erroneous results, it is capable of predicting many nonlinear effects, including limit cycles and the 'jump resonance' phenomenon, which arises when the frequency-response becomes multivalued. Moreover, it can be extended, by the inclusion of more frequencies, to cover other effects such as the generation of subharmonics and the quenching of limit cycles. This approach is sometimes referred to as 'harmonic linearisation', since it constitutes another example of the general methodology whereby nonlinear systems are studied, as far as possible, using linear techniques.

2 STATE-SPACE MODELS

The essential feature of the concept of 'state' for a dynamical system is that it should contain all information about the past history of the system which is relevant to its future behaviour. That is to say, if the state at a given instant is known, then its subsequent evolution can be predicted without any other knowledge of what has previously happened to the system. In the context of a description by ordinary differential equations, this means that the definition of the state must involve enough variables for all the equations to be of first order in the time-derivative. The dependent variables in these equations are called the state variables, and we denote them in general by $x_1,x_2,\ldots,x_n$, where n is known as the order of the system. Since there is one equation for each state variable, the order is thus the number of independent equations and hence also the number of independent initial conditions. Besides the state variables, however, the equations will also contain a number of externally specified variables which constitute the 'driving forces' acting upon the system from outside. These are known as input variables, and we shall denote them by $u_1,u_2,\ldots,u_m$, where m is usually, though not necessarily, less then n (often very much less). The differential equations can thus be written in the form

$$\begin{aligned}
\dot{x}_1 &= f_1(x_1,\ldots,x_n,\, u_1,\ldots,u_m,t)\\
\dot{x}_2 &= f_2(x_1,\ldots,x_n,\, u_1,\ldots,u_m,t)\\
&\vdots\\
\dot{x}_n &= f_n(x_1,\ldots,x_n,\, u_1,\ldots,u_m,t)
\end{aligned}$$

where $f_1,f_2,\ldots,f_n$ are in general nonlinear functions. For ease of notation, we assemble the state and input variables into vectors

$$\begin{aligned}
\mathbf{x} &= (x_1,x_2,\ldots,x_n)^{\mathrm{T}}\\
\mathbf{u} &= (u_1,u_2,\ldots,u_m)^{\mathrm{T}}
\end{aligned}$$

where the superscript T denotes that the transpose is to be taken, that is to say, **x** and **u** are actually column vectors although written in row form to save space. With a similar notation for the functions on the right-hand side, the equations then become

$$\dot{\mathbf{x}} = \mathbf{f}(\mathbf{x},\mathbf{u},t)$$

which is a vector differential equation to be solved for $\mathbf{x}(t)$, given $\mathbf{u}(t)$ for $t \geqslant t_0$ and the initial state $\mathbf{x}(t_0)$. The solution of this equation gives complete information about the behaviour of the system, but we may only be interested in certain aspects, represented by a set of variables which are functions of $\mathbf{x}$,$\mathbf{u}$ and t. These are called output variables and are denoted by $y_1,y_2,\ldots,y_p$, where p is usually less than n; they may be, for example, a subset of the state variables. In general, we have

$$\begin{aligned} y_1 &= h_1(\mathbf{x},\mathbf{u},t) \\ y_2 &= h_2(\mathbf{x},\mathbf{u},t) \\ &\vdots \\ y_p &= h_p(\mathbf{x},\mathbf{u}.t) \end{aligned}$$

where $h_1,h_2,\ldots,h_p$ may be nonlinear functions. Defining the output vector

$$\mathbf{y} = (y_1,y_2,\ldots,y_p)^{\mathrm{T}}$$

we can then write

$$\mathbf{y} = \mathbf{h}(\mathbf{x},\mathbf{u},t)$$

where the scalar functions on the right-hand side have been collected into a single vector function.

2.1 Existence and uniqueness

Since the state vector is supposed to contain sufficient information about the system at any given time for its subsequent behaviour to be predicted, it is necessary that the differential equation for $\mathbf{x}(t)$ should have a unique solution for every initial state $\mathbf{x}(t_0)$ and input vector $\mathbf{u}(t)$. This can be ensured by imposing some constraints on the permissible form of the vector function $\mathbf{f}(\mathbf{x},\mathbf{u},t)$, or rather, on the functions

$$\mathbf{f}_{\mathbf{u}}(\mathbf{x},t) \equiv \mathbf{f}(\mathbf{x},\mathbf{u}(t),t)$$

which result from the insertion of any particular input. For a solution to exist, it is sufficient that $\mathbf{f}_{\mathbf{u}}(\mathbf{x},t)$ should be continuous in all its arguments, but this does not by itself guarantee uniqueness. In order to ensure this, it is usual to impose a further restriction, called a Lipschitz condition, for which we need to introduce a vector norm, such as

$$|\mathbf{x}| = \sqrt{(x_1^2 + x_2^2 + \ldots + x_n^2)}$$

which is the Euclidean length of the vector $\mathbf{x}$. The Lipschitz condition then takes the form

$$|\mathbf{f_u}(\mathbf{x},t) - \mathbf{f_u}(\tilde{\mathbf{x}},t)| \leqslant K|\mathbf{x}-\tilde{\mathbf{x}}|$$

for all $\mathbf{x}$ and $\tilde{\mathbf{x}}$ in a specified region, and all t in some interval, where K is a constant. This condition, of course, automatically implies that $\mathbf{f}$ is continuous with respect to $\mathbf{x}$ and, combined with continuity in t, it guarantees both existence and uniqueness of solutions.

The above conditions are, however, somewhat stronger than we would like. To begin with, the assumption of continuity with respect to t would rule out discontinuous input signals, such as step functions, which are often used in specifying the performance of control systems. Actually, this is not a very serious difficulty since, provided that $\mathbf{f_u}(\mathbf{x},t)$ has only a finite number of discontinuities in any interval, we can simply split up the time axis at the corresponding points, construct the solutions in the zones thus created and join them at the boundaries. On the other hand, it is also often useful to allow discontinuities with respect to the state variables, and this requires rather more care, since it entails violation of the Lipschitz condition and hence possibly loss of uniqueness. To see what may happen in such a case, suppose $\mathbf{f_u}(\mathbf{x},t)$ is discontinuous across some hypersurface, as illustrated in Fig. 2.1. This means that, immediately on either side of the discontinuity, $\dot{\mathbf{x}}$ will take different values, which we can denote by $\dot{\mathbf{x}}_a$, $\dot{\mathbf{x}}_b$, corresponding to the sides labelled a,b, respectively. Then, letting $\mathbf{v}$ denote the unit vector normal to the hypersurface, directed from side a to side b, we have the following possibilities:

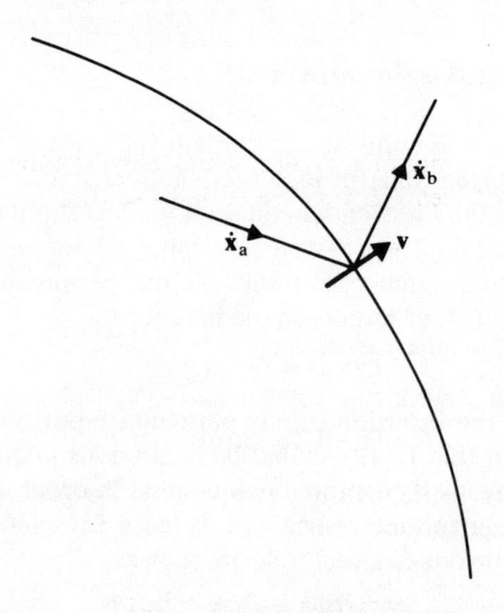

Fig. 2.1 Discontinuity in $\dot{\mathbf{x}}$ across a hypersurface.

(i) $\mathbf{v}^{\mathrm{T}}\dot{\mathbf{x}}_{\mathrm{a}} > 0$, $\mathbf{v}^{\mathrm{T}}\dot{\mathbf{x}}_{\mathrm{b}} > 0$ or $\mathbf{v}^{\mathrm{T}}\dot{\mathbf{x}}_{\mathrm{a}} < 0$, $\mathbf{v}^{\mathrm{T}}\dot{\mathbf{x}}_{\mathrm{b}} < 0$. In this case, there is no problem, because the equations can be integrated until the discontinuity is reached on one side, and then continued on the other, using the new expression for $\dot{\mathbf{x}}$.

(ii) $\mathbf{v}^{\mathrm{T}}\dot{\mathbf{x}}_{\mathrm{a}} > 0$, $\mathbf{v}^{\mathrm{T}}\dot{\mathbf{x}}_{\mathrm{b}} < 0$. This means that the representative point approaches the discontinuity from each side, with the natural implication that it will then remain on the hypersurface and that a more precise specification of $\mathbf{f}_{\mathbf{u}}(\mathbf{x},t)$ is needed to predict its subsequent motion.

(iii) $\mathbf{v}^{\mathrm{T}}\dot{\mathbf{x}}_{\mathrm{a}} < 0$, $\mathbf{v}^{\mathrm{T}}\dot{\mathbf{x}}_{\mathrm{b}} > 0$. Here we have a genuine ambiguity, which arises because the system is inadequately modelled and can only be avoided by modifying the equations so as to eliminate this type of discontinuity in $\mathbf{v}^{\mathrm{T}}\dot{\mathbf{x}}$.

(iv) $\mathbf{v}^{\mathrm{T}}\dot{\mathbf{x}}_{\mathrm{a}} = 0$ or $\mathbf{v}^{\mathrm{T}}\dot{\mathbf{x}}_{\mathrm{b}} = 0$ or both. If this happens, the outcome is unclear and a more detailed investigation will be required.

2.2 Linearisation

In view of the much greater simplicity and tractability of linear, as opposed to nonlinear, equations, it is obviously desirable, whenever possible, to make use of linear approximations. For this approach to be valid, we require that the nonlinear functions $\mathbf{f}(\mathbf{x},\mathbf{u},t)$ and $\mathbf{h}(\mathbf{x},\mathbf{u},t)$ are smooth enough to be approximated by linear Taylor expansions with respect to the components of $\mathbf{x}$ and $\mathbf{u}$. That is to say, if we write

$$\begin{aligned} \mathbf{u} &= \hat{\mathbf{u}} + \Delta\mathbf{u} \\ \mathbf{x} &= \hat{\mathbf{x}} + \Delta\mathbf{x} \\ \mathbf{y} &= \hat{\mathbf{y}} + \Delta\mathbf{y} \end{aligned}$$

where $\hat{\mathbf{u}}$, $\hat{\mathbf{x}}$ and $\hat{\mathbf{y}}$ are nominal values, which may be functions of t, we can define gradient matrices $\nabla_{\mathbf{x}}\mathbf{f}$, $\nabla_{\mathbf{u}}\mathbf{f}$, $\nabla_{\mathbf{x}}\mathbf{h}$, $\nabla_{\mathbf{u}}\mathbf{h}$, by setting

$$(\nabla_{\mathbf{x}}\mathbf{f})_{jk} = \frac{\partial f_j}{\partial x_k}$$

and similarly for the others, such that

$$\frac{|\mathbf{f}(\mathbf{x},\mathbf{u},t) - \mathbf{f}(\hat{\mathbf{x}},\hat{\mathbf{u}},t) - \nabla_{\mathbf{x}}\mathbf{f}(\hat{\mathbf{x}},\hat{\mathbf{u}},t)\Delta\mathbf{x} - \nabla_{\mathbf{u}}\mathbf{f}(\hat{\mathbf{x}},\hat{\mathbf{u}},t)\Delta\mathbf{u}|}{\max(|\Delta\mathbf{x}|, |\Delta\mathbf{u}|)} \to 0$$

as $\Delta\mathbf{x} \to 0$ and $\Delta\mathbf{u} \to 0$, with a corresponding result for $\mathbf{h}$. We can then obtain the linearised model

$$\begin{aligned} \frac{\mathrm{d}}{\mathrm{d}t}(\Delta\mathbf{x}) &= \mathbf{A}\Delta\mathbf{x} + \mathbf{B}\Delta\mathbf{u} \\ \Delta\mathbf{y} &= \mathbf{C}\Delta\mathbf{x} + \mathbf{D}\Delta\mathbf{u} \end{aligned}$$

where the coefficient matrices are given by

$$\mathbf{A} = \nabla_{\mathbf{x}}\mathbf{f}(\hat{\mathbf{x}},\hat{\mathbf{u}},t) \qquad \mathbf{B} = \nabla_{\mathbf{u}}\mathbf{f}(\hat{\mathbf{x}},\hat{\mathbf{u}},t)$$
$$\mathbf{C} = \nabla_{\mathbf{x}}\mathbf{h}(\hat{\mathbf{x}},\hat{\mathbf{u}},t) \qquad \mathbf{D} = \nabla_{\mathbf{u}}\mathbf{h}(\hat{\mathbf{x}},\hat{\mathbf{u}},t)$$

and the nominal functions $\hat{\mathbf{x}}(t)$, $\hat{\mathbf{u}}(t)$ and $\hat{\mathbf{y}}(t)$ satisfy

$$\frac{d\hat{\mathbf{x}}}{dt} = \mathbf{f}(\hat{\mathbf{x}},\hat{\mathbf{u}},t) \qquad \hat{\mathbf{y}} = \mathbf{h}(\hat{\mathbf{x}},\hat{\mathbf{u}},t)$$

representing the solution around which the equations have been linearised. The linear model may in general be time-dependent, even if the original system was not explicitly so. However, if $\hat{\mathbf{u}}$ is held constant and $\hat{\mathbf{x}}$ is an equilibrium point, so that

$$\mathbf{f}(\hat{\mathbf{x}},\hat{\mathbf{u}}) = 0, \qquad \hat{\mathbf{y}} = \mathbf{h}(\hat{\mathbf{x}},\hat{\mathbf{u}})$$

where we assume that the nonlinear system is time-invariant, then so is the linearised version.

2.3 Autonomous systems

Although the equations of a dynamical model will in general depend on the time, either explicitly or through the input function $\mathbf{u}(t)$, or both, a large part of nonlinear system theory is concerned with cases where there is no time dependence at all. Such systems are said to be autonomous, and they arise quite naturally in practice when, for example, the input vector is held fixed, as in the case of a control system where it might be a constant reference or demand signal. Indeed, much more complicated input signals, such as polynomials, exponentials and sinusoids, can also be included in this category, by regarding them as generated by auxiliary dynamical systems with constant inputs, which are then incorporated into the complete model. In any such case, the differential equation for the state vector will become

$$\dot{\mathbf{x}} = \mathbf{f}(\mathbf{x},\hat{\mathbf{u}})$$

where $\hat{\mathbf{u}}$ is a constant vector which is, however, to be regarded as adjustable, since we may wish to know how its value affects the behaviour of the system. Thus, the equilibrium points in the state-space, that is to say, the values $\hat{\mathbf{x}}$ at which $\mathbf{x}(t)$ would remain fixed if initially placed there, are given by

$$\mathbf{f}(\hat{\mathbf{x}},\hat{\mathbf{u}}) = 0$$

whose solutions (if any) will generally depend, both in value and multiplicity, on $\hat{\mathbf{u}}$. Further, if the model is then linearised about a particular equilibrium condition given by $\hat{\mathbf{x}}$ and $\hat{\mathbf{u}}$, the coefficient matrices of the resulting linear system will depend upon the values chosen for these vectors.

Assuming that $\mathbf{f}(\mathbf{x},\hat{\mathbf{u}})$ satisfies a Lipschitz condition, the differential equation for $\mathbf{x}(t)$ will have a unique solution, for any given initial state $\mathbf{x}(0)$. The path traced out in the state space by $\mathbf{x}(t)$ is called a trajectory of the system and, because of the uniqueness property, there will be one and only one trajectory passing through any given point. In the case of an equilibrium point, the corresponding trajectory will, of course, degenerate into the point itself, which is therefore sometimes known as a singular point.

2.4 The phase portrait

If we suppress the dependence on $\hat{\mathbf{u}}$, which amounts to regarding the input vector as fixed, the state-space differential equations for an autonomous system can be written simply as

$$\dot{\mathbf{x}} = \mathbf{f}(\mathbf{x})$$

and the set of all trajectories of this equation provides a complete geometrical representation of the dynamical behaviour of the system, under the specified conditions. This is sometimes referred to as the phase portrait although, strictly speaking, this term derives from a particular type of state-space representation in which the equations take the form

$$\begin{aligned} \dot{x}_1 &= x_2 \\ \dot{x}_2 &= x_3 \\ &\vdots \\ \dot{x}_{n-1} &= x_n \\ \dot{x}_n &= \phi(x_1,x_2,\ldots,x_n) \end{aligned}$$

where all the nontrivial functional dependence is contained in ϕ, and the components of the state vector are called ‘phase variables’. Nevertheless, since no radically new feature is introduced by the use of this representation, we shall continue to employ the same terminology whether the equations are actually in phase-variable form or not. This pictorial description of the system’s behaviour is mainly of use for second-order systems, where the state–space becomes a plane, called the ‘phase plane’. For higher-order cases, there are obvious difficulties of visualisation, but there is also an essential increase in complication, at least potentially. It arises because the fact that the trajectories cannot cross (due to uniqueness) imposes particular restrictions on the possible structure of a two-dimensional phase portrait, which do not apply in general. Consequently, and also because many practically important systems can be approximately represented by using only two state variables, such as position and velocity, the majority of analytical effort has been directed to this case. As a result, it is possible to give an essentially complete classification of behaviour in the phase plane, though not in higher-dimensional state spaces.

2.5 Singular points

The equilibrium points of an autonomous system, given by

$$\mathbf{f}(\hat{\mathbf{x}}) = 0$$

are also known as singular points, because they appear to violate the general rule that only one trajectory can pass through any given point. Actually, the violation is only apparent, since the trajectories which meet at a singular point do not really pass through it, but only approach or depart from it asymptotically, depending on the stability properties of the equilibrium condition. Assuming that $\mathbf{f}(\mathbf{x})$ is smooth enough for the equations to be linearised around the singular point $\hat{\mathbf{x}}$, the linear approximation is given by

$$\frac{\mathrm{d}}{\mathrm{d}t}(\Delta\mathbf{x}) = \mathbf{A}\Delta\mathbf{x}$$

where

$$\mathbf{A} = \nabla_{\mathbf{x}}\mathbf{f}(\hat{\mathbf{x}})$$

and, provided that $\mathbf{A}$ is nonsingular, this approximation will be sufficient to determine the behaviour of the trajectories in the neighbourhood of the equilibrium point. The nonsingularity condition is necessary, since otherwise the linear terms might not be dominant and could even vanish entirely; also, the singular point might not be isolated. If, however, $\mathbf{A}$ is nonsingular, the nature of the equilibrium is essentially determined by its eigenvalues and can be classified accordingly. Thus, the equilibrium point is sometimes said to be of 'type m' if $\mathbf{A}$ has m eigenvalues with positive real parts, so that a stable point is of type 0, and all those of higher type are unstable. For second-order systems, there is a more detailed conventional classification, as follows:

'stable node' eigenvalues real and negative;
'stable focus' eigenvalues complex conjugate, with negative real part;
'unstable node' eigenvalues real and positive;
'unstable focus' eigenvalues complex conjugate, with positive real part;
'saddle point' eigenvalues real and of opposite sign;
'centre' eigenvalues conjugate pure imaginary.

These are illustrated in Figs. 2.2 to 2.7.

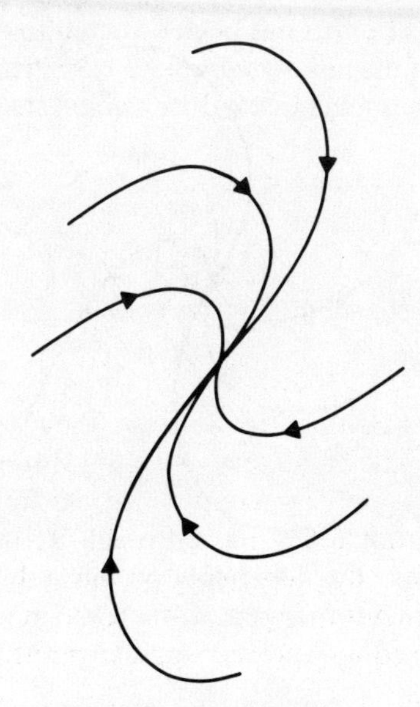

Fig. 2.2 Stable node.

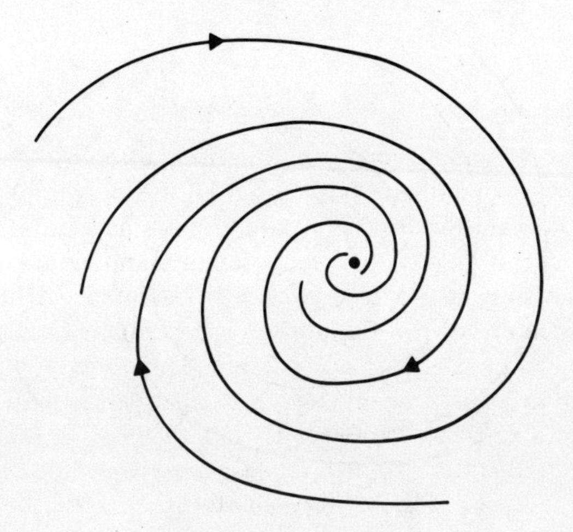

Fig. 2.3 Stable focus.

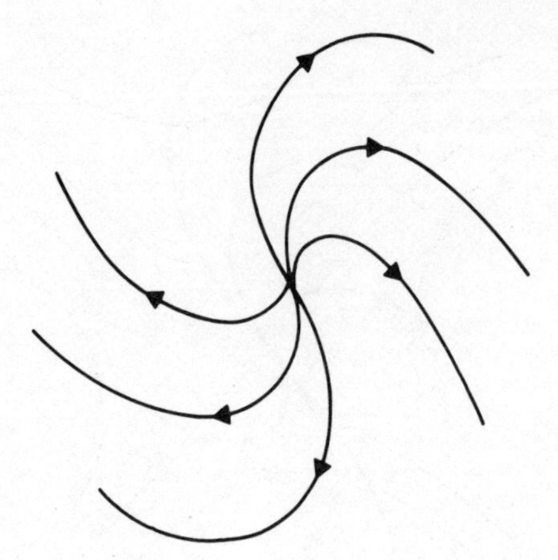

Fig. 2.4 Unstable node.

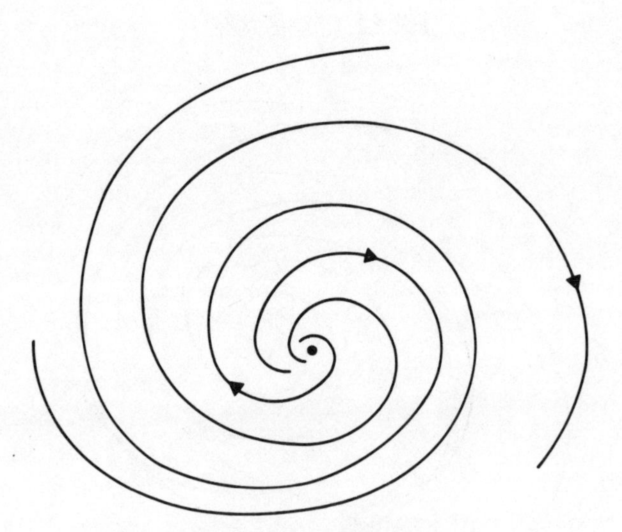

Fig. 2.5 Unstable focus.

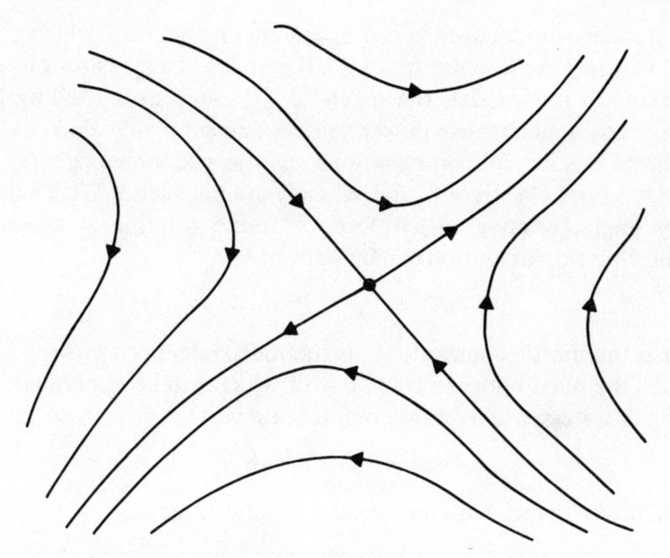

Fig. 2.6 Saddle point.

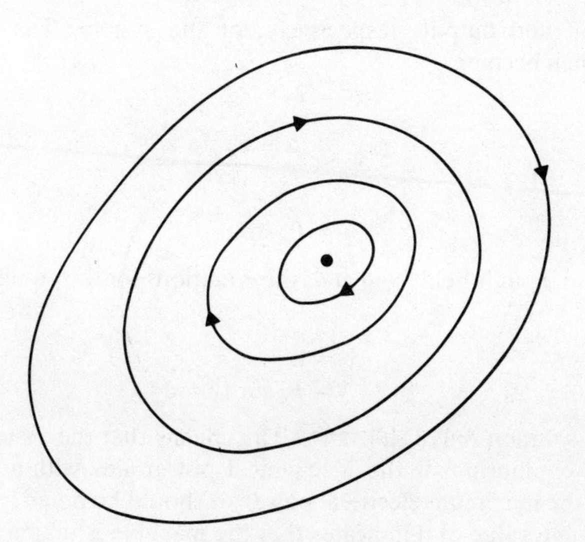

Fig. 2.7 Centre.

Example 2.1 The swing equations for a synchronous machine

This is a standard model used in power system engineering, which gives the simplest plausible representation of the dynamics of a synchronous electrical machine. In this model, the machine is treated as a rotating inertial mass, driven by a mechanical power source, and supplying electrical power to an infinite busbar; all complications such as friction, magnetic saturation, variation of field strength and saliency are neglected. With δ denoting the rotor angle, relative to a reference frame rotating at synchronous speed, the equation of motion can be written as

$$H\ddot{\delta} = P_m - P_e \sin \delta$$

where H is the inertia constant, P_m is the mechanical power supplied, and P_e denotes the maximum electrical power which can be generated. To put the model in state-space form, we define state variables

$$x_1 = \delta \qquad x_2 = \dot{\delta}$$

and also, in accordance with our general notation, set

$$u = P_m$$
$$y = P_e \sin \delta$$

as the input and output, respectively, of the system. The state-space equations then become

$$\dot{x}_1 = x_2$$

$$\dot{x}_2 = \frac{u - P_e \sin x_1}{H}$$

$$y = P_e \sin x_1$$

so that, if the input is held fixed at $\hat{u}$, the equations for the equilibrium state $\hat{x}$ are

$$\hat{x}_2 = 0$$
$$\hat{y} = P_e \sin \hat{x}_1 = \hat{u}$$

with a real solution only if $|\hat{u}| \leqslant P_e$. This means that the system can only operate in equilibrium if the mechanical power lies within the bounds defined by the maximum electrical power; it should be noted, incidentally, that a negative value of u indicates that the machine is taking power from the bus and performing mechanical work, in other words, acting as a motor instead of a generator. Assuming that the equilibrium equations can be satisfied, we see that there are in fact two essentially distinct solutions, given by

$$\hat{x}_1 = \arcsin\left(\frac{\hat{u}}{P_e}\right)$$

$$\hat{x}_1 = \pi - \arcsin\left(\frac{\hat{u}}{P_e}\right)$$

respectively; of course, there are also an infinite number of other solutions, obtained by adding multiples of 2π radians, but these correspond to the same physical conditions as those above.

If we now linearise the equations about one of the equilibrium points, we obtain a set of equations in the standard linear state-space form, with coefficient matrices

$$\mathbf{A} = \begin{bmatrix} 0 & 1 \\ \dfrac{-P_e \cos \hat{x}_1}{H} & 0 \end{bmatrix}, \quad \mathbf{B} = \begin{bmatrix} 0 \\ \dfrac{1}{H} \end{bmatrix},$$

$$\mathbf{C} = [\ P_e \cos \hat{x}_1 \quad 0\ ], \quad \mathbf{D} = 0,$$

giving the transfer function

$$G(s) = \frac{P_e \cos \hat{x}_1}{Hs^2 + P_e \cos \hat{x}_1}$$

from which it is apparent that the behaviour of the linearised system depends strongly on which equilibrium point has been chosen. The first one gives a 'marginally stable' resonant model, with poles on the imaginary axis; the other has both poles real, one of them positive, and is therefore unstable.

Returning to the nonlinear system, but keeping $u = \hat{u}$, it is easy to show that the system is conservative, in the sense that there exists a scalar function

$$V(x) = \tfrac{1}{2} H x_2^2 - \hat{u} x_1 - P_e \cos x_1$$

such that the state-space equations give

$$\dot{V} = 0$$

and so the trajectories are obtained by setting V to constant values, that is to say, they are contours of $V(x)$, as illustrated in Fig. 2.8. From the linearised equations, we see that the stable singular points are centres and the unstable ones are saddle points. An interesting special case is given by setting $\hat{u} = 0$, when the equations become equivalent to those of a simple pendulum, with $V(x)$ as the total energy. In practice, the system will not be quite conservative because of dissipative effects, which may be crudely modelled by adding a viscous damping term $K\dot{\delta}$ to the left-hand side of the equation of motion, giving

$$\dot{V} = -K\dot{\delta}^2 = -Kx_2^2$$

with the result that, although the singular points are unaltered in position, the stable ones now become foci (asymptotically stable), as shown in Fig. 2.9.

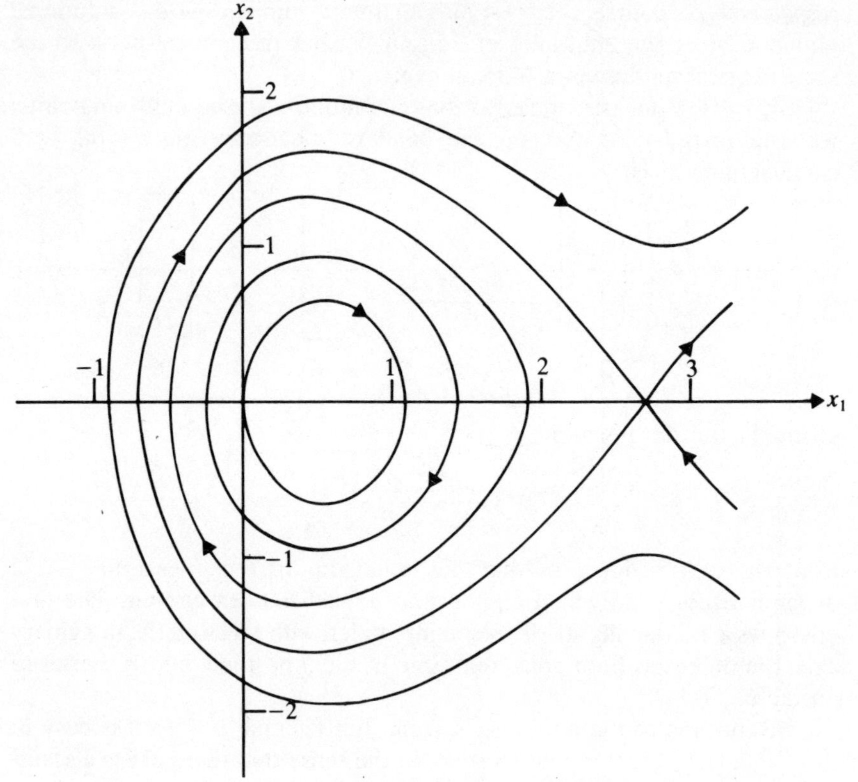

Fig. 2.8 Trajectories for undamped swing equations ($H = 1$, $P_m = 1$, $P_e = 2$).

2.6 Isoclines

In general, the equations describing a nonlinear system cannot be solved analytically, so that, in order to construct the trajectories accurately, it is necessary to use numerical methods. With the help of the powerful algorithms now available for this purpose, one can simulate the behaviour of the system on a digital computer, display the results graphically, and hence build up the phase portrait by beginning the simulation from different points. This, however, may be expensive in computer time, and still fail to reveal important features, unless the initial conditions are sufficiently varied and well chosen. Consequently, it is still worthwhile to obtain as much information as possible by analytical means, before

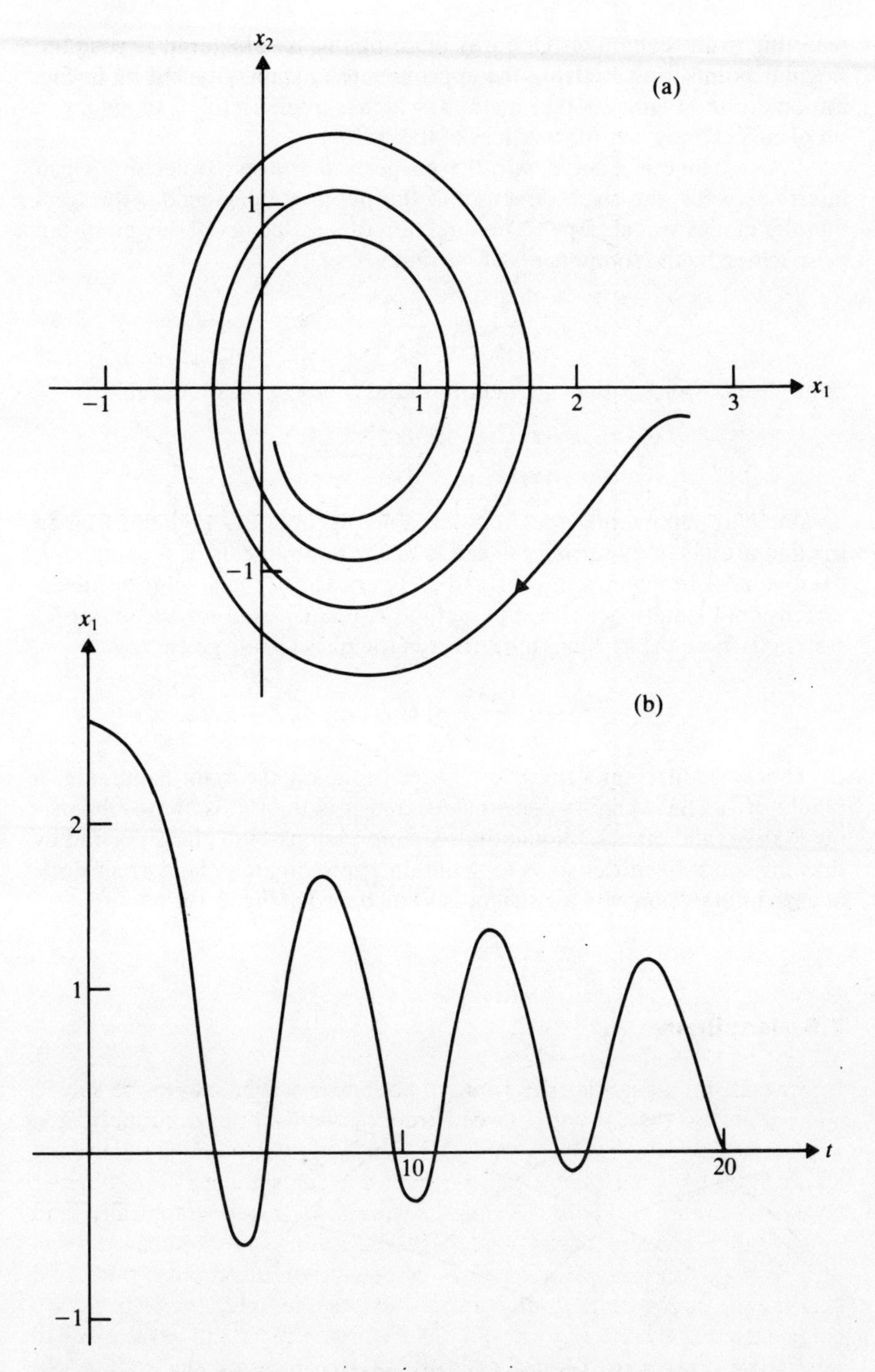

Fig. 2.9(a) Trajectory for swing equations with damping ($K = 0.1$); (b) corresponding plot of $\delta(t)$.

resorting to the computer. One way of doing this is to begin by finding the singular points, and studying the approximate models obtained by linearisation around them; another method, which is often useful, is to employ a set of curves known as the isoclines of the system.

An isocline is a curve with the property that every trajectory which intersects it has the same direction at the point of intersection; the term literally means 'equal slope'. The direction of a trajectory at any point can be specified by the components of the unit vector

$$\frac{\dot{\mathbf{x}}}{|\dot{\mathbf{x}}|} = \frac{\mathbf{f}(\mathbf{x})}{|\mathbf{f}(\mathbf{x})|}$$

or by a set of $(n-1)$ independent differential coefficients, for example

$$\frac{dx_2}{dx_1} = \frac{f_2(\mathbf{x})}{f_1(\mathbf{x})}, \ldots, \frac{dx_n}{dx_1} = \frac{f_n(\mathbf{x})}{f_1(\mathbf{x})}$$

obtained by eliminating the time variable. Hence, the equations of an isocline are given by setting the ratios of the components of $\mathbf{f}(\mathbf{x})$ equal to fixed values. In general, this gives $(n-1)$ equations for a curve in the n-dimensional state-space, but the method is mainly used for second-order systems, where the isoclines are curves in the phase plane given by

$$\frac{f_2(\mathbf{x})}{f_1(\mathbf{x})} = \text{constant.}$$

By choosing different values for the constant on the right-hand side, a family of isoclines can be constructed, and it is then possible to obtain a qualitative (and often semiquantitative) impression of the phase portrait by drawing the trajectories so as to maintain approximately the correct slope at each intersection with an isocline, as illustrated in Fig. 2.10.

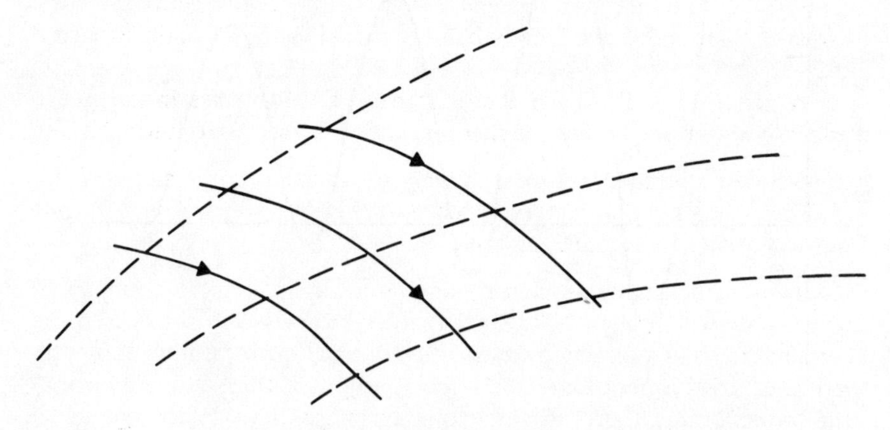

Fig. 2.10 Isoclines (dashed) and trajectories (solid).

2.7 Limit cycles

A common feature of autonomous systems is the occurrence of a special type of trajectory which takes the form of a closed curve. This is known as a limit cycle and represents a periodic solution of the system equations since, when the state vector returns to its initial value, it must necessarily repeat its previous motion and so continue indefinitely. The components of $\mathbf{x}(t)$ are thus oscillatory functions of time, so that, as the system is autonomous, the limit cycle represents an oscillation which is intrinsic to the system and not imposed from outside. In a rather trivial sense, this description could include the trajectories of certain linear systems, such as a simple harmonic oscillator, but its use is normally confined to nonlinear cases where the limit cycle is an isolated phenomenon; that is to say, other trajectories in its vicinity are not limit cycles. We can, in fact, classify the stability properties in much the same way as for equilibrium points, regarding a limit cycle as stable if nearby trajectories approach it asymptotically and unstable if they move away. The analysis, however, is now much more difficult since the linearised equations are time-dependent, through the periodic nature of the limit cycle solution itself. In practice, only stable limit cycles can actually be observed as persistent oscillations of a system, just as only stable equilibria can constitute permanent points of rest, owing to the inevitable presence of disturbances.

Limit cycles can occur in systems of any order, and indeed constitute the typical form of oscillatory behaviour which arises when an equilibrium point of a nonlinear system becomes unstable, as a result, for example, of excessive gain being applied in a control loop. They are, nevertheless, particularly characteristic of second-order systems, since a closed curve is the only possible limiting form of a phase-plane trajectory as $t \to \infty$, apart from an asymptotic approach to a singular point or to infinity. Consequently, and also because the analysis is simpler, conditions for their existence or absence have mainly been obtained in the second-order case only.

A general condition for the existence of limit cycles in the phase plane is the Poincaré-Bendixson theorem, which states the following:

> if a closed bounded region contains no singular point and has the property that some trajectory which starts inside never leaves it, then it must contain at least one limit cycle.

This theorem is usually applied by finding a region, typically annular in shape as shown in Fig. 2.11, such that all trajectories which cross its boundary do so in a direction towards the interior of the region. It follows that trajectories cannot leave the region and hence that, if it contains no equilibrium point, there must be one or more limit cycles within it; moreover, at least one will be stable.

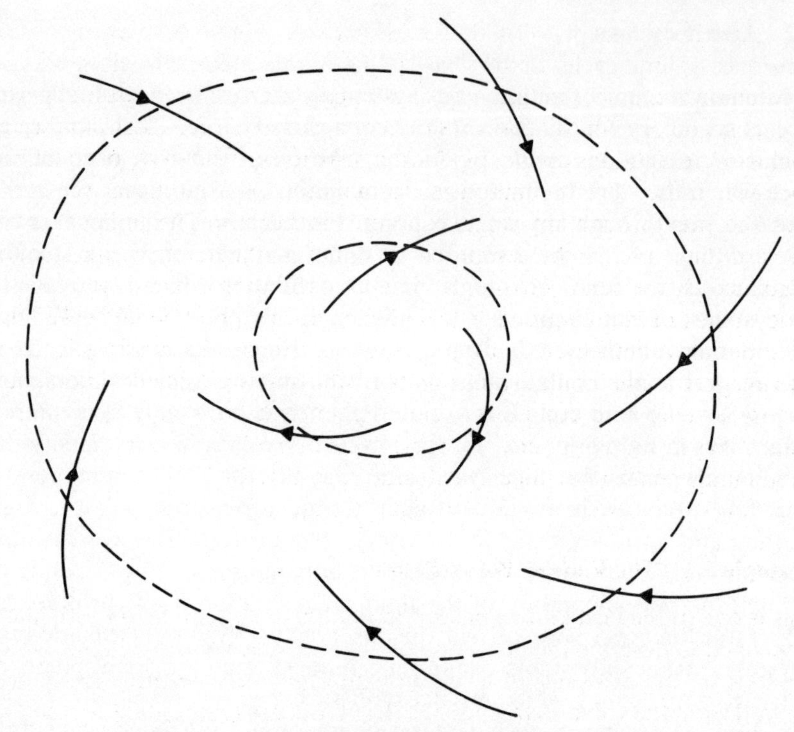

Fig. 2.11 Use of the Poincaré–Bendixson theorem.

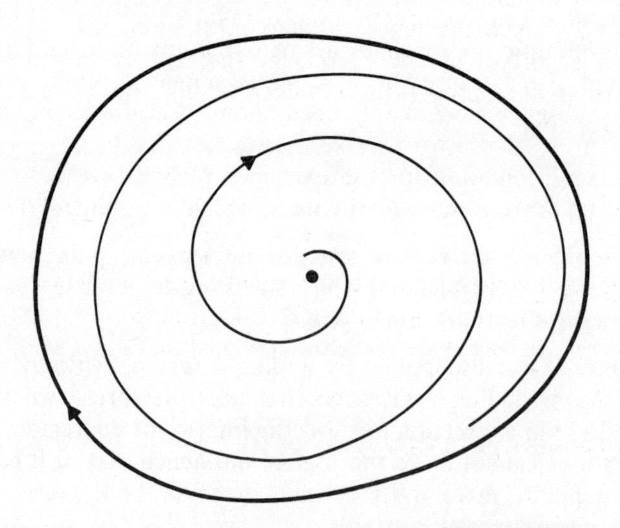

Fig. 2.12 Limit cycle surrounding unstable focus (Poincaré index = +1).

Another concept, also due to Poincaré, which is relevant to the occurence of limit cycles in the phase plane, is the 'index' of a closed curve. If the curve is simple, that is to say, it does not intersect itself, its index with respect to the vector function $\mathbf{f}(\mathbf{x})$ is defined as the net total number of clockwise revolutions made by $\mathbf{f}$ as $\mathbf{x}$ traverses the curve once in the clockwise sense. This definition is unambiguous provided that the curve does not pass through any singular point. Furthermore, it implies that the index of the curve can be computed by summing the 'contributions' of the singular points which it surrounds, assuming they are isolated, where each node, focus, or centre counts $+1$, and each saddle point counts -1. Since the index of a limit cycle is clearly $+1$, this restricts its possible location with respect to the equilibrium points of the system. Specifically, it must enclose at least one equilibrium point, and if there is only one, then it cannot be a saddle point. A typical case, with a stable limit cycle surrounding an unstable focus, is illustrated in Fig. 2.12.

Example 2.2 The Van der Pol oscillator

This is one of the best known models in nonlinear system theory, originally developed to describe the operation of an electronic valve oscillator, which depends on the existence of a region with effectively negative resistance, derived from the valve characteristic. The basic equation is that of a simple harmonic oscillator with nonlinear damping, which can be written, after appropriate scaling of the time variable, as

$$\ddot{w} + \phi(\dot{w}) + w = 0$$

where ϕ is a nonlinear function which has negative gradient in some interval, and w is a dynamic variable; in the original application, w represented a current and $\dot{w}$ a voltage. In fact, the equation is often presented in an alternative form, obtained by differentiating with respect to time and setting

$$y = \dot{w}$$

which gives

$$\ddot{y} + \phi'(y)\dot{y} + y = 0$$

but this has no obvious advantage, so we will return to the original version. Since the qualitative results do not depend strongly on the exact form of the function ϕ, we may as well take the simplest (and usual) choice

$$\phi(\dot{w}) = \epsilon(\dot{w}^3 - \dot{w})$$

where ϵ is a positive parameter. Taking

$$x_1 = w \qquad x_2 = \dot{w}$$

as state variables, we then have

$$\dot{x}_1 = x_2$$
$$\dot{x}_2 = -x_1 + \epsilon(x_2 - x_2^3)$$

for the state-space equations. The system evidently has only one equilibrium point, namely the origin, which is shown by linearisation to be unstable; it is a focus for $\epsilon < 2$, a node for $\epsilon \geqslant 2$. Some further guidance as to the shape of the phase portrait can be obtained by constructing the isoclines. Thus, setting

$$\frac{dx_2}{dx_1} = R$$

we get

$$x_1 = (\epsilon - R)\, x_2 - \epsilon x_2^3$$

for the isocline corresponding to slope R. With the help of the isoclines, drawn for a few suitably chosen values of R, a sketch of the phase portrait can be constructed as in Fig. 2.13. This clearly indicates the presence of a stable limit cycle, which is asymptotically approached as $t \to \infty$ by every trajectory, except for the degenerate case $\mathbf{x}(t) \equiv 0$. The existence of the limit cycle can be rigorously established by applying the Poincaré-Bendixson theorem to an appropriate annular region; any circle of radius

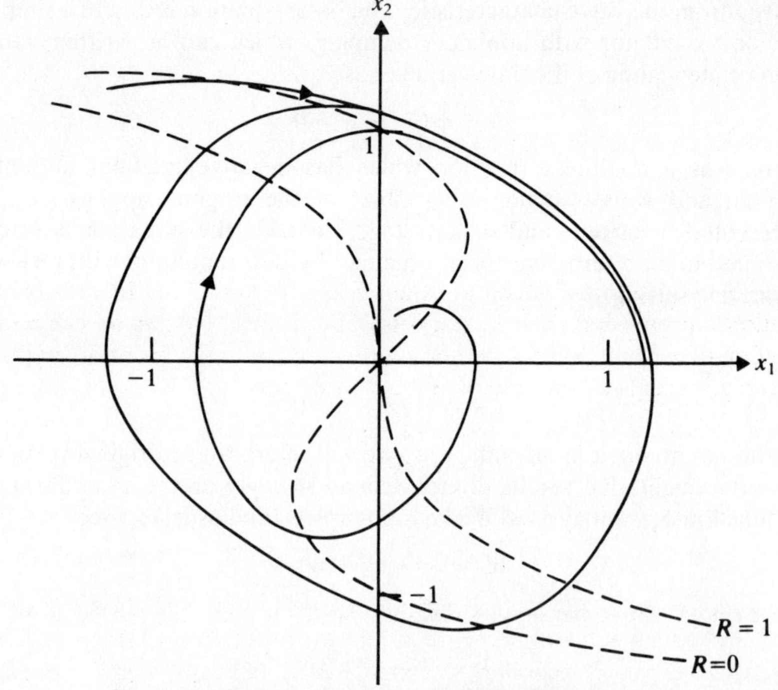

Fig. 2.13 Trajectories of Van der Pol oscillator ($\epsilon = 1$).

less than unity, centred at the origin, will serve as the inner boundary, though some care is needed in defining the outer one.

Example 2.3 The Volterra–Lotka ecosystem model

Although it has applications in other areas, such as chemical reaction kinetics, this type of model is usually associated with the description of an ecological system comprising several competing species. In its simplest form, it contains just two species, one of which preys upon the other. Denoting the populations (or, more legitimately, the biomasses) of the prey and predator species, respectively, by x_1 and x_2, which we treat as continuous variables, we assume that the logarithmic growth rate of each species is dependent only on the current population of the other, so that

$$\frac{\dot{x}_1}{x_1} = \phi_2(x_2)$$

$$\frac{\dot{x}_2}{x_2} = \phi_1(x_1)$$

where ϕ_1 and ϕ_2 are appropriately chosen functions. Somewhat surprisingly at first sight, this is a conservative system since, if we define

$$V(\mathbf{x}) = \psi_1(x_1) - \psi_2(x_2)$$

where

$$\frac{\partial \psi_j}{\partial x_j} = \frac{\phi_j}{x_j}$$

for $j = 1,2$, it follows that

$$\dot{V} = 0$$

along the trajectories, which are therefore contours of V. In view of the relationship between the species, it is natural to take ϕ_1 as an increasing function of x_1, and ϕ_2 as a decreasing function of x_2; for want of anything better, it is usual to choose linear functions, giving

$$\dot{x}_1 = (a - bx_2)\, x_1$$
$$\dot{x}_2 = (cx_1 - d)\, x_2$$

where a,b,c,d are positive constants. The conserved quantity then becomes

$$V = cx_1 - d \ln x_1 + bx_2 - a \ln x_2$$

whose contours are the trajectories, illustrated in Fig. 2.14. In the positive quadrant, which is the only region of interest for this application, they are closed curves, corresponding to an oscillatory behaviour of the populations. It appears from the phase portrait that there are two singular points,

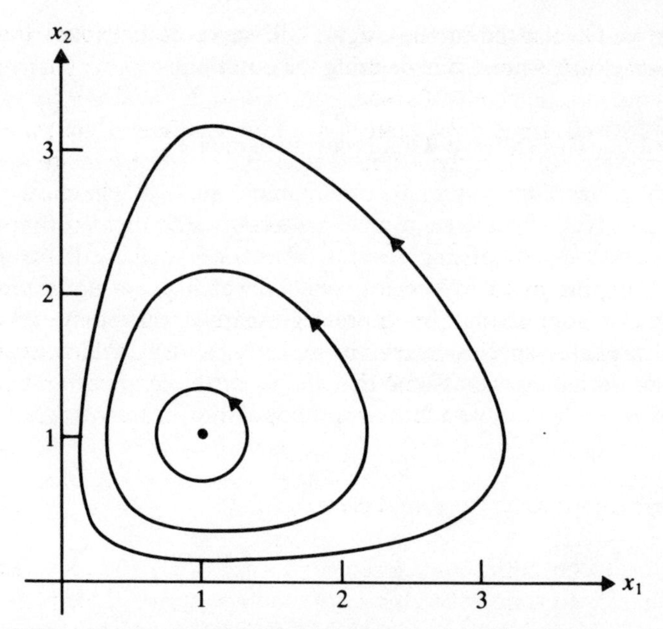

Fig. 2.14 Trajectories of Volterra-Lotka equations ($a = b = c = d = 1$).

whose locations are readily obtained from the system equations; they are at (0,0) and (d/c, a/b). Linearisation about the origin is trivial and shows it to be a saddle point; for the other equilibrium point, the plant matrix of the linearised system is

$$\mathbf{A} = \begin{pmatrix} 0 & -bd/c \\ ac/b & 0 \end{pmatrix}$$

whose eigenvalues are pure imaginary, so that the point is a centre, as would be expected from the undampened nature of the oscillations.

The applicability of this model to any real ecosystem is dubious, to say the least, even though natural populations are sometimes observed to exhibit roughly periodic fluctuations. It is nevertheless interesting to explore the possibilities of modifying the model so as to incorporate some other effects which one might expect to be present in reality. For example, in the model as it stands, there is nothing, apart from predation, to prevent the prey species from expanding indefinitely, which is hardly realistic. In order to include self-limiting effects, we could try altering the differential equation for the prey population to

$$\dot{x}_1 = ax_1 - bx_1x_2 - \mu x_1^2$$

for some positive μ, where the extra term represents the result, for instance, of overcrowding and disease. This has the effect of damping the oscillations, so that the stable equilibrium point becomes a focus, instead of a centre. On the other hand, at least in the equation for the predator population, there should arise a delay, since the effect of a change in food supply is not instantaneous. To model this, we could replace $x_1(t)$ by $x_1(t-\tau)$, though this would make the state-space infinite-dimensional; however, if the delay were small, it would be reasonable to approximate $x_1(t-\tau)$ by $(x_1-\tau\dot{x}_1)$, giving

$$\dot{x}_2 = cx_1x_2 - dx_2 - c\tau x_2\dot{x}_1$$

where $\dot{x}_1$ is given by the other state equation. This tends to destabilise the system and, for certain parameter values, the result can be that the singular point is unstable while large oscillations are still damped, so that a limit cycle appears, as shown in Fig. 2.15.

2.8 Strange attractors and chaos

Although singular points and closed curves constitute the only asymptotic forms of bounded trajectories for autonomous systems in the phase plane, this is no longer true in spaces of higher dimension. In general, we define the 'positive limit set' of a trajectory $\mathbf{x}(t)$ as the set of all points $\mathbf{q}$ for which, given any $\epsilon > 0$, there exists a sequence of instants t_k, with

$$t_k \to \infty$$

as $k \to \infty$, such that

$$|\mathbf{q}-\mathbf{x}(t_k)| < \epsilon$$

for every positive integer k. If the trajectory is bounded, that is to say, there is a constant μ such that

$$|\mathbf{x}(t)| < \mu$$

for all $t > 0$, then the positive limit set Ω cannot be empty. This is proved by invoking the Bolzano–Weierstrass theorem of functional analysis, which states that a bounded infinite sequence of points always contains a convergent subsequence. In the same way, it follows that the trajectory asymptotically approaches Ω, in the sense that

$$\inf_{\mathbf{q}} |\mathbf{q} - \mathbf{x}(t)| \to 0$$

as $t \to \infty$, where the infimum is taken over all $\mathbf{q}$ in Ω. It is also easily shown that Ω is closed, bounded and connected.

We should note at this point, however, that the foregoing analysis does not imply anything about stability. Thus, an unstable singular point or limit cycle automatically qualifies as the positive limit set of a particular trajectory, namely itself, just as well as if it were stable. In order to discuss

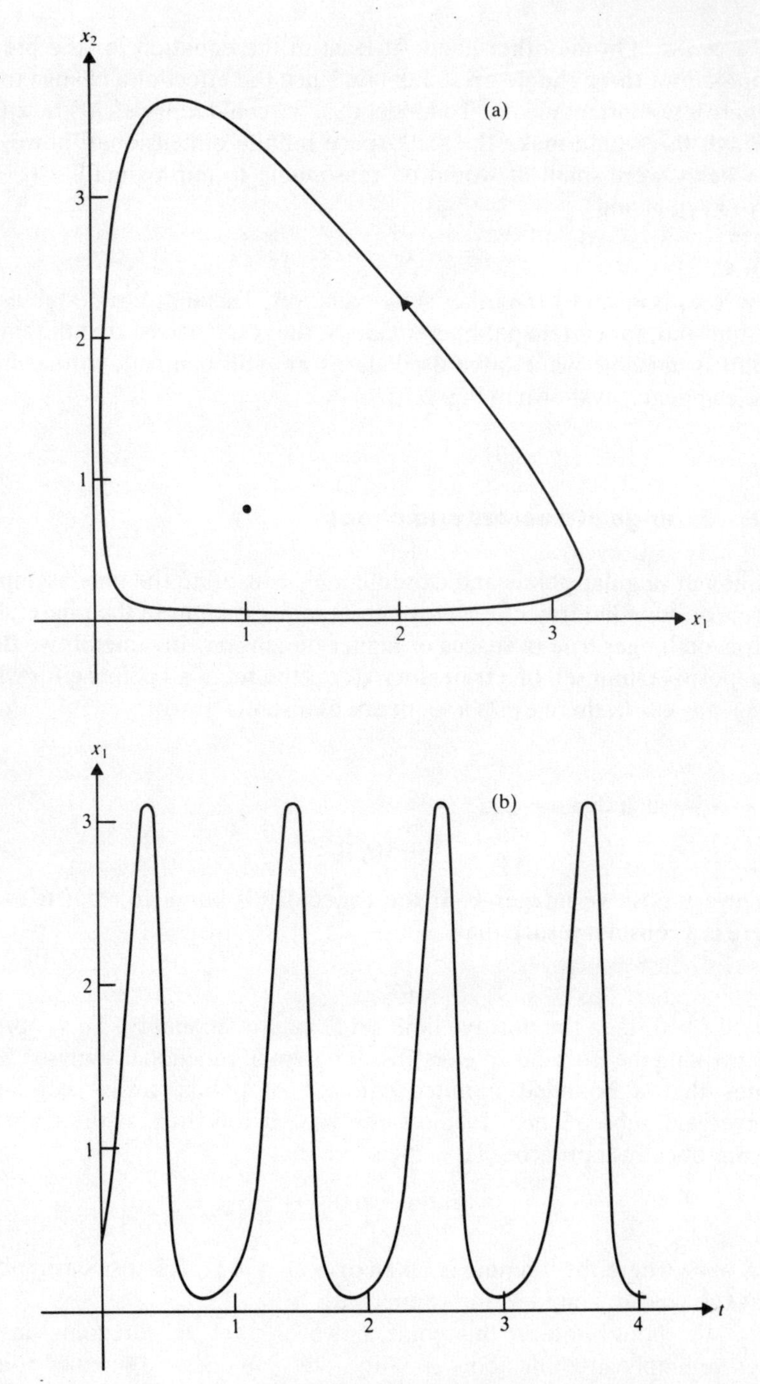

Fig. 2.15(a) Limit cycle in modified Volterra–Lotka system ($\mu = 0.2$, $\tau = 0.3$); (b) corresponding plot of $x_1\ (t)$.

stability, we need to consider the behaviour of neighbouring trajectories, as when we define an equilibrium to be (asymptotically) stable, or a limit cycle 'orbitally stable', if all trajectories in its vicinity approach it as $t \to \infty$. The general term for a set with this property is an 'attractor', since it asymptotically attracts nearby trajectories to itself. Also, the phrase 'limit set' may be understood as including both positive and negative cases, where the 'negative limit set' of a trajectory is defined as for the positive case, except that $t_k \to -\infty$ instead of ∞.

For second-order systems, the only types of limit set normally encountered are singular points and limit cycles. There is another theoretical possibility, namely a closed curve consisting of one or more complete trajectories terminated by singular points, as in Fig. 2.16, but it appears to be rather uncommon in practice. In a state-space of more than two dimensions, however, a far greater variety of behaviour is possible. Even for linear systems, there are some essentially new features, such as the occurrence of almost-periodic solutions in a system consisting of two or more harmonic oscillators with incommensurable frequencies. This clearly requires a state space of at least four dimensions, and shows the existence of limit sets which are sections of hypersurfaces, rather than curves. A similar phenomenon can appear in nonlinear systems, even with only three dimensions, as exemplified by the 'Kronecker flow', illustrated in Fig. 2.17, where the limit set is a torus, on which an infinite number of trajectories approach arbitrarily close to every point without intersecting one another. More complicated still are the so-called 'strange' limit sets, which have been found in many nonlinear systems of order greater than two and could even be typical of them. These may or may not be asymptotically attractive to neighbouring trajectories; if so, they are known as 'strange attractors', though even then, the trajectories they contain may be locally divergent from each other, within the attracting set. Such structures are associated with the quasi-random behaviour of solutions called 'chaos', although the connection between them is not altogether clear. It appears that there may

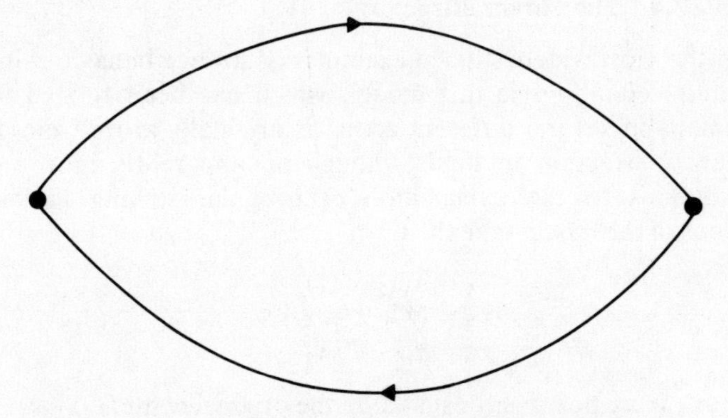

Fig. 2.16 Closed curve formed by two complete trajectories.

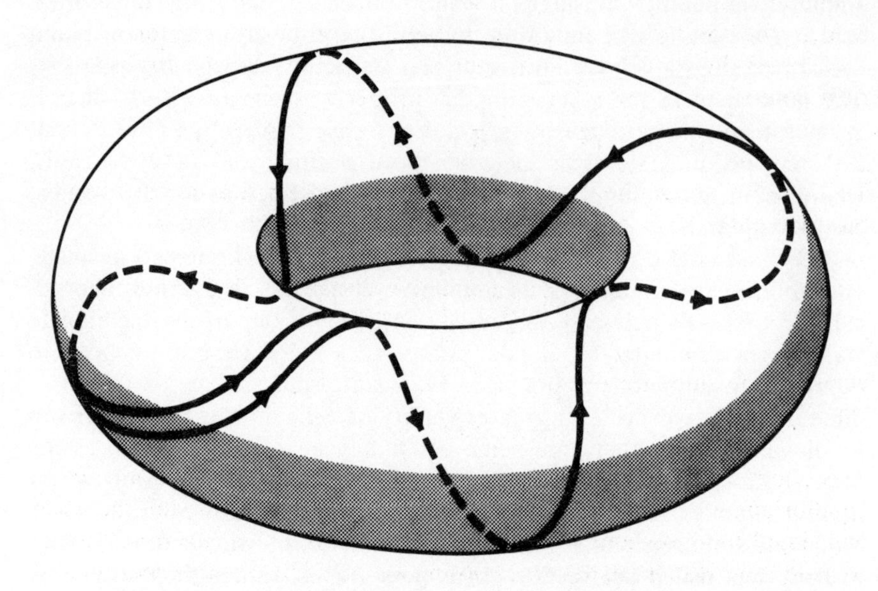

Fig. 2.17 The Kronecker flow on a torus.

be different types of chaotic behaviour, depending, for instance, on whether the corresponding strange limit set is an attractor or not. Much of what is known about this topic has been obtained by simulation studies, which makes it difficult to be precise about the exact nature of the trajectories. Indeed, one cannot even be sure, from a finite-length simulation, that a solution is not periodic (of sufficiently long period), let alone whether or not it is almost-periodic, although the conclusion may become highly plausible with enough data.

Example 2.4 The Lorenz attractor

One of the most widely studied examples of strange behaviour in ordinary differential equations is this model, which has been related to chaotic phenomena in several different areas. It originally arose from a study of turbulent convection in fluids, though an apparently more convincing application is to the explanation of irregular spiking in lasers. The equations of the model take the form

$$\begin{aligned} \dot{x}_1 &= \sigma(x_2 - x_1) \\ \dot{x}_2 &= (1 + \lambda - x_3)x_1 - x_2 \\ \dot{x}_3 &= x_1 x_2 - b x_3 \end{aligned}$$

where σ,λ,b are positive constants. In the original context, x_1 was a Fourier component of the fluid velocity field, x_2 and x_3 were similarly related to the temperate gradient, while in the laser application, x_1 represents the electric field, x_2 the polarisation and $(1 + \lambda - x_3)$ the population inversion.

From the state-space equations, we find that there are three equilibrium points:

$$(0,0,0), \quad (\sqrt{b\lambda}, \sqrt{b\lambda},\lambda), \quad (-\sqrt{b\lambda},-\sqrt{b\lambda},\lambda).$$

To linearise about the origin, we can simply drop the terms involving products of the state variables, so that the plant matrix becomes

$$\begin{pmatrix} -\sigma & \sigma & 0 \\ \lambda+1 & -1 & 0 \\ 0 & 0 & -b \end{pmatrix}$$

whose eigenvalues are given by

$$(s+b)\,\{s^2+(\sigma+1)s-\sigma\lambda\} = 0$$

This equation clearly has all its roots real, one being positive, and so the equilibrium is unstable. Because of the symmetry of the model, the other two singular points have effectively identical properties, so we need only consider one of them in detail. Choosing the one with positive coordinates, we obtain a linearised model with plant matrix

$$\begin{pmatrix} -\sigma & \sigma & 0 \\ 1 & -1 & -\sqrt{b\lambda} \\ \sqrt{b\lambda} & \sqrt{b\lambda} & -b \end{pmatrix}$$

for which the eigenvalue equation is

$$s^3+(\sigma+b+1)s^2 + b(\sigma+\lambda+1)s + 2\sigma b\lambda = 0.$$

For this equation to have all its roots in the open left half-plane, the Routh–Hurwitz theorem of classical stability theory gives the necessary and sufficient condition

$$(\sigma+b+1)\,b(\sigma+\lambda+1) > 2\sigma b\lambda$$

which can be rewritten as

$$\lambda(b+1-\sigma) + (\sigma+1)\,(\sigma+b+1) > 0.$$

Thus, when this condition is satisfied, both equilibrium points, away from the origin, are asymptotically stable. On the other hand, if we have

$$\sigma > b + 1$$

$$\lambda > \frac{(\sigma+1)\,(\sigma+b+1)}{\sigma-b-1}$$

then all the equilibria are unstable. Under these circumstances, the system exhibits very complicated behaviour, in which a typical trajectory, starting near one of the equilibrium points, spirals gradually outwards for some time and then moves towards the 'opposite' equilibrium, around which it executes a similar performance, and so on. Projections of such a trajectory on the (x_1,x_2) and (x_2,x_3) planes are shown in Figs 2.18 and 2.19, respectively, and a plot of $x_1(t)$ in Fig. 2.20. This kind of bounded but irregular motion, consisting of alternating sequences containing randomly varying numbers of oscillations, is a characteristic indication that a strange attractor is present. Also, in this particular system, it appears from simulation studies that, for certain parameter ranges, the strange attractor can persist even though the equilibrium asymptotic stability condition holds, so that both types of behaviour coexist.

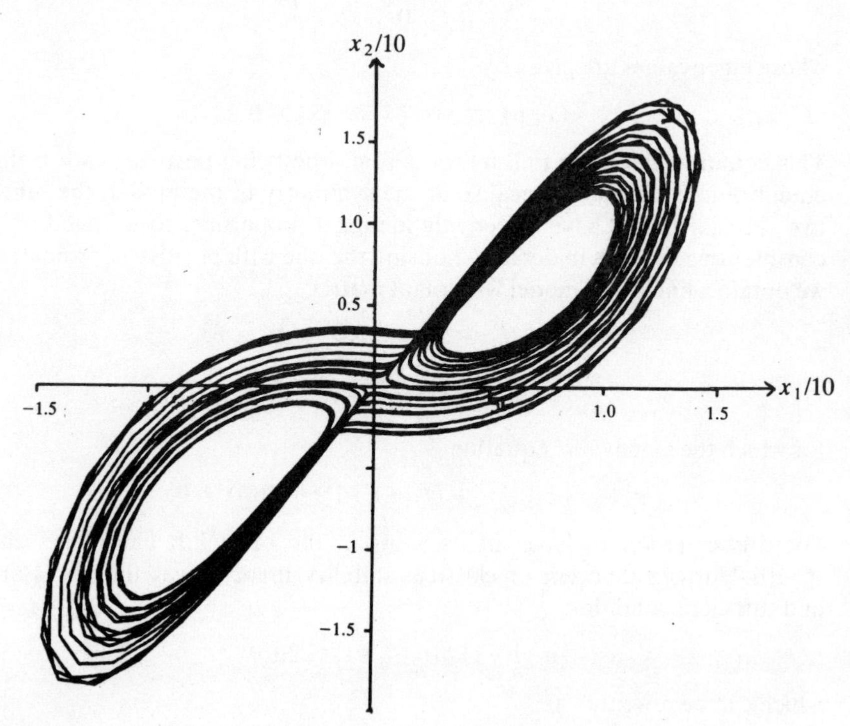

Fig. 2.18 Projection on the (x_1,x_2) plane of a trajectory of the Lorenz equations ($\sigma = 10, \lambda = 24, b = 2$).

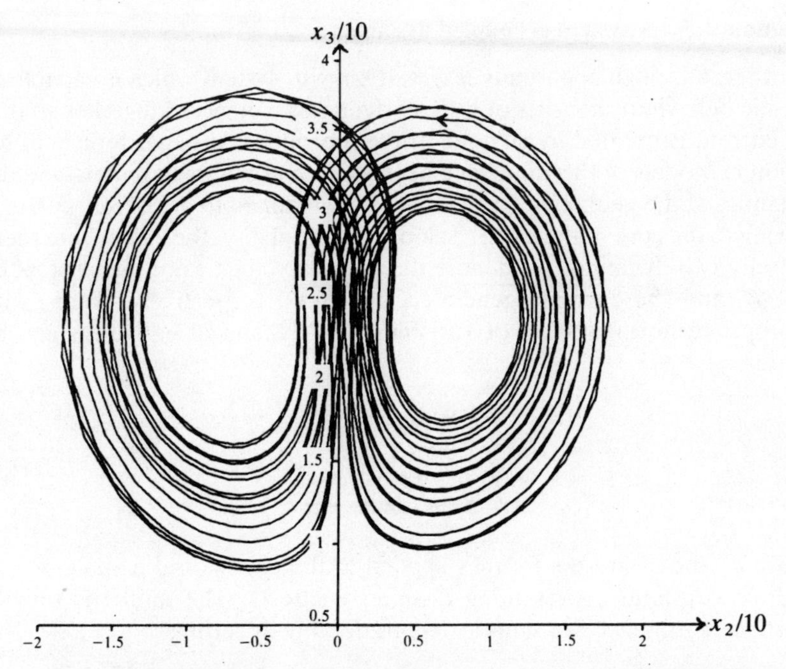

Fig. 2.19 Projection on the (x_2, x_3) plane for trajectory of Fig. 2.18.

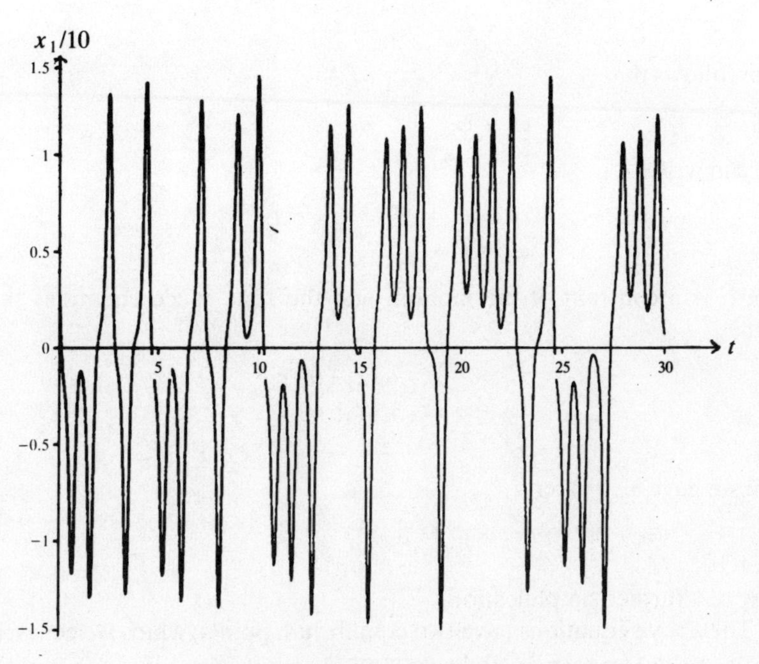

Fig. 2.20 Plot of $x_1(t)$ for Fig. 2.18.

Example 2.5 A system of coupled dynamos

Another, although apparently less well known, system which is capable of chaotic behaviour, consists of a set of dynamos connected together so that the current generated by any one of them produces the magnetic field for another. Models of this kind have been used in attempts to understand the dynamics of the geomagnetic field, which is assumed to arise from currents flowing in the core of the earth. Taking, for simplicity, the case where there are only two dynamos, we denote the angular velocities of their rotors by ω_1, ω_2, and the currents generated by x_1, x_2, respectively. Then, with appropriate normalisation of variables, the dynamical equations can be written

$$\begin{aligned}
\dot{x}_1 &= -\mu_1 x_1 + \omega_1 x_2 \\
\dot{x}_2 &= -\mu_2 x_2 + \omega_2 x_1 \\
\dot{\omega}_1 &= q_1 - \epsilon_1 \omega_1 - x_1 x_2 \\
\dot{\omega}_2 &= q_2 - \epsilon_2 \omega_2 - x_1 x_2
\end{aligned}$$

where q_1 and q_2 are the torques applied to the rotors, and $\mu_1,\mu_2,\epsilon_1,\epsilon_2$ are positive constants representing dissipative effects. The model is thus of fourth order, but we can simplify it considerably by setting

$$q_1 = q_2 = 1$$

and making the rather artificial assumption that

$$\epsilon_1 = \epsilon_2 = 0.$$

It then follows that

$$\dot{\omega}_1 = \dot{\omega}_2$$

so we can write

$$\begin{aligned}
\omega_1 &= x_3 + \alpha \\
\omega_2 &= x_3 - \alpha
\end{aligned}$$

where α is a constant of the motion, and the state-space equations thus become

$$\begin{aligned}
\dot{x}_1 &= -\mu x_1 + (x_3+\alpha)\, x_2 \\
\dot{x}_2 &= -\mu x_2 + (x_3-\alpha)\, x_1 \\
\dot{x}_3 &= 1 - x_1 x_2
\end{aligned}$$

where we have also taken

$$\mu_1 = \mu_2 = \mu$$

by way of a further simplification.

The above equations have two equilibrium points, which we can write as (β_1,β_2,γ) and $(-\beta_1,-\beta_2,\gamma)$, by defining

$$\beta_1 = \sqrt{\left(\frac{\gamma+\alpha}{\mu}\right)}$$

$$\beta_2 = \sqrt{\left(\frac{\gamma-\alpha}{\mu}\right)}$$

$$\gamma = \sqrt{(\alpha^2+\mu^2)}.$$

Since the model is unaltered by reversing the signs of x_1 and x_2, both singular points have the same dynamical properties. Linearising around the one in the positive orthant, we obtain the plant matrix

$$\begin{pmatrix} -\mu & \gamma+\alpha & \beta_2 \\ \gamma-\alpha & -\mu & \beta_1 \\ -\beta_2 & -\beta_1 & 0 \end{pmatrix}$$

whose eigenvalue equation factorises as

$$(s+2\mu)\left(s^2+\frac{2\gamma}{\mu}\right)=0$$

so that the linearised system is 'marginally' stable, with a pair of pure imaginary eigenvalues. To discover the behaviour of trajectories away from the singular points, it appears necessary to resort to simulation methods, and the results turn out to depend, as might be expected, on the values of μ and α. For some parameter values, a limit cycle exists, though of a rather distorted and highly non-planar form, as shown by the projections in Figs 2.21 and 2.22. With other choices of the parameters, however, the system develops the sort of behaviour typical of a strange attractor, as illustrated in Fig. 2.23 by projection on the (x_1,x_3) plane, resulting in the plot of $x_3(t)$ shown in Fig. 2.24.

2.9 Exercises

1 With viscous damping included, the synchronous machine equation in Example 2.1 becomes

$$H\ddot{\delta} + K\dot{\delta} = P_m - P_e \sin\delta$$

where H,K,P_e are positive constants. Linearise the state equations around a fixed value of P_m, for both the stable and unstable equilibrium points, and calculate the transfer function from ΔP_m to $\Delta\delta$ in each case.

2 For the Van der Pol oscillator of Example 2.2, construct a region in the phase plane, to which the Poincaré–Bendixson theorem can be applied, and hence prove the existence of a limit cycle.

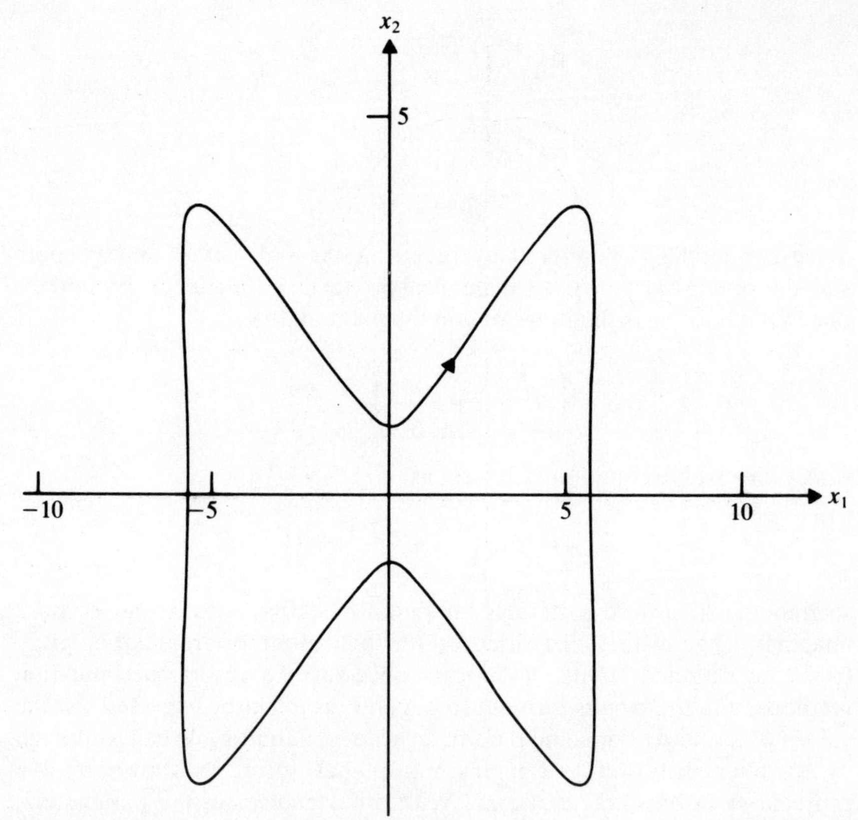

Fig. 2.21 Projection on the (x_1,x_2) plane of a limit cycle in the coupled dynamo model ($\mu = 0.2, \alpha = 1$).

3 After the modifications suggested in Example 2.3, the Volterra–Lotka equations, with $a=b=c=d=1$, become

$$\dot{x}_1 = x_1 - x_1x_2 - \mu x_1^2$$
$$\dot{x}_2 = x_1x_2 - x_2 - \tau x_2\dot{x}_1.$$

Find the equilibrium points, linearise about the one with nonzero values of both state variables, and show how its stability or instability depends on μ and τ. Discuss the possibility of a limit cycle appearing.

4 Solve the system of equations

$$\dot{x}_1 = \upsilon x_2 - x_1x_3$$
$$\dot{x}_2 = -\upsilon x_1 - x_2x_3$$
$$\dot{x}_3 = \ln \sqrt{(x_1^2 + x_2^2)}$$

and show that the trajectories lie on toroidal surfaces. Consider how the results depend on the parameter υ.

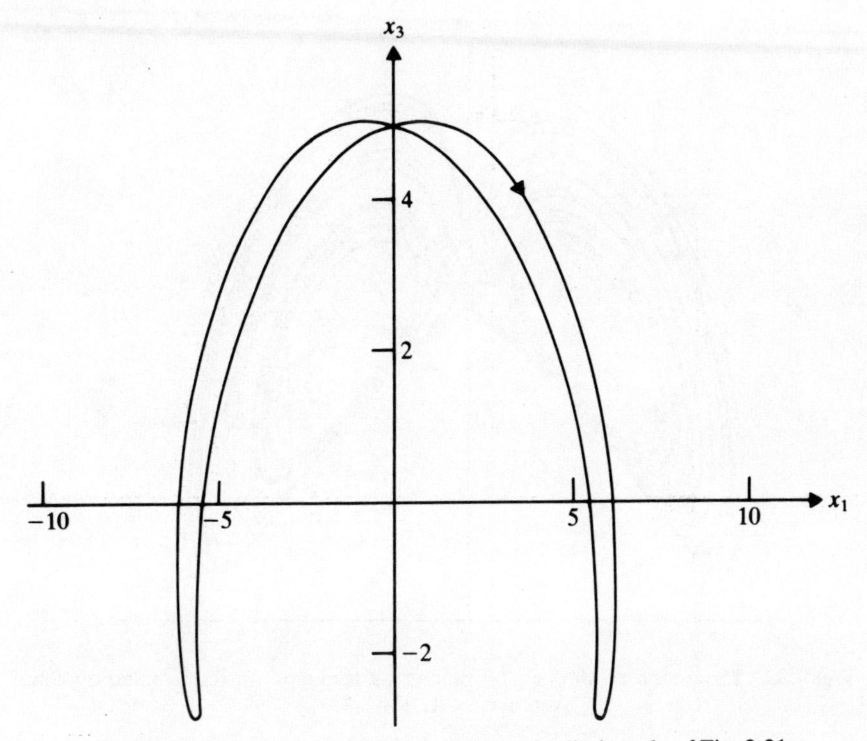

Fig. 2.22 Projection on the (x_1, x_3) plane for limit cycle of Fig. 2.21.

5 Investigate the Lorenz equations of Example 2.4 when the positivity constraints on the constant parameters are removed. Show how the numbers of stable and unstable equilibria depend on the values of σ, λ and b.

6 A chaotic model due to Rössler has the equations

$$\dot{x}_1 = x_1 - x_1 x_2 - x_3$$
$$\dot{x}_2 = x_1^2 - a x_2$$
$$\dot{x}_3 = b x_1 - c x_3$$

for constant a, b and c. Locate the singular points of the system, linearise the equations around them, and discuss their stability as a function of the parameters.

7 In the coupled dynamo model of Example 2.5, investigate the effect of taking $q_1 \neq q_2$ and

$$\epsilon_1 = \epsilon_2 = \epsilon > 0.$$

Find how the positions and types of equilibria vary as ϵ is increased from 0 to ∞, with q_1 and q_2 fixed.

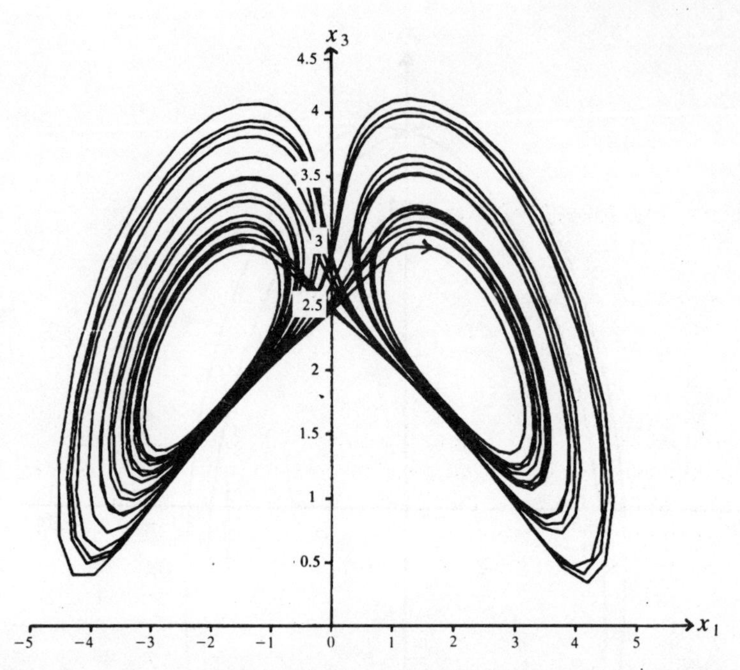

Fig. 2.23 Projection on the (x_1, x_3) plane of a trajectory in the coupled dynamo system ($\mu = 1, \alpha = 1.875$).

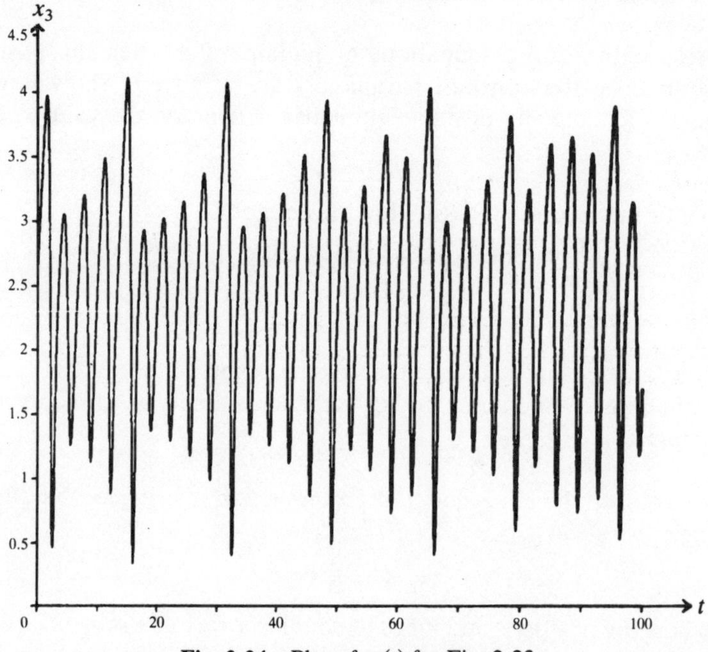

Fig. 2.24 Plot of $x_3(t)$ for Fig. 2.23.

3 HARMONIC ANALYSIS

One of the commonest techniques in dynamical system analysis is the exploitation, in one way or another, of periodicity in time. For linear time-invariant systems, this is particularly simple and leads naturally to the familiar frequency-domain description, via Laplace or Fourier transformation. In the nonlinear case, on the other hand, although the simplicity is lost, the technique is no less important, since periodic phenomena, such as limit cycles, are among the most prominent features of many systems. Consequently, much attention has been given to methods of understanding and predicting such effects, even if only in an approximate sense.

The natural framework for the analysis of periodic behaviour is the Fourier series, so we will begin with a brief review of its properties. Suppose we have a scalar function $y(t)$, which is periodic with period T, that is to say,

$$y(t+T) = y(t)$$

for all t. Since the simplest periodic functions are constants and sinusoids, we attempt to represent $y(t)$ by a series of the form

$$y(t) = c_0 + \sum_{k=1}^{\infty} \{a_k \cos(k\omega t) + b_k \sin(k\omega t)\}$$

where ω (the angular frequency) has the value $\omega = 2\pi/T$ in order to give the correct periodicity. For this representation to be valid, the coefficients must be given by the expressions

$$a_k = \frac{1}{\pi}\int_0^{2\pi} y(t)\cos(k\omega t)\,\mathrm{d}(\omega t)$$

$$b_k = \frac{1}{\pi}\int_0^{2\pi} y(t)\sin(k\omega t)\,\mathrm{d}(\omega t)$$

$$c_0 = \frac{1}{2\pi}\int_0^{2\pi} y(t)\,\mathrm{d}(\omega t) = \tfrac{1}{2}a_0$$

which are obtained by substituting the series expansion for $y(t)$ and integrating term-by-term. Assuming that $y(t)$ is piecewise-continuous, the series can be shown to converge to the correct value at every point where $y(t)$ is continuous; moreover, at any point of discontinuity, it converges to the average of the two values obtained by taking the limit of $y(t)$ as t approaches the point from each side. The proof of these properties involves another result (the Riemann-Lebesgue lemma) which states that a_k and b_k both tend to zero as $k \to \infty$, and this is of considerable importance since it helps to justify the approximation of $y(t)$ by truncating the series after a finite number of terms. This procedure can also be easily shown to generate the best possible approximation of $y(t)$ in a certain sense (mean-square approximation) since, if we define an approximant

$$y_n(t) = c_0 + \sum_{k=1}^{n} \{a_k \cos(k\omega t) + b_k \sin(k\omega t)\}$$

and minimise the squared-error integral

$$\int_0^{2\pi} \{y(t) - y_n(t)\}^2 \, d(\omega t)$$

with respect to the coefficients, regarded as free parameters, we obtain the same values for a_k, b_k, c_0 as were given above.

For the purposes of nonlinear system analysis, the importance of the Fourier series arises from two considerations. First, it provides a decomposition of a general periodic signal into a set of terms, each of which retains its own form (a sinusoid of a particular frequency) under the operation of any linear system element. Second, it lends itself naturally to the construction of an approximation scheme by truncation of the series. Taken together, these properties suggest a technique for the approximate analysis of periodic phenomena, which goes by the name of 'harmonic balance' or the 'describing function method', where every dynamical variable is approximated by a finite sum of periodic terms, usually consisting of a single-frequency (purely sinusoidal) oscillation with possibly a constant bias. Any nonlinear component of the system is then approximately represented by its action on sinusoids just as though it were linear, so that the method involves a kind of quasi-linearisation. It should be noted, also, that this approach can be generalised somewhat beyond the strictly periodic case, since the foregoing considerations do not depend on the terms in the Fourier series all being at multiples of the same fundamental frequency. In fact, a more general class of 'almost-periodic' functions can be defined by a 'generalised Fourier series' in which the terms have arbitrary frequencies, of the form

$$y(t) = c_0 + \sum_{k=1}^{\infty} \{a_k \cos(\omega_k t) + b_k \sin(\omega_k t)\}$$

where the ω_k may be incommensurable. This suggests an extension of the method to cover cases where signals are approximated by a combination of sinusoidal terms at unrelated frequencies, which is naturally appropriate for the study of oscillations in systems driven by external inputs of periodic (or almost-periodic) form.

3.1 Describing functions

The basic idea of this method is to represent a nonlinear system element by a kind of 'transfer function', derived from its effects on sinusoidal input signals or combinations of these. Considering a scalar nonlinearity with input $u(t)$ and output $y(t)$, the simplest input function of this type is

$$u(t) = U \sin(\omega t)$$

where U and ω are positive constants. Assuming that the nonlinearity is an 'instantaneous' function, with no internal dynamics, the output will then also be a periodic function (of period $2\pi/\omega$), which can be expanded in a Fourier series as above. For certain classes of nonlinear input–output relation, we can make some deductions about the coefficients in the series, as follows:

(i) if y is an odd function of u (odd symmetry), then changing t to $t + (\pi/\omega)$, and hence u to $-u$, changes y to $-y$, which implies that

$$a_k = b_k = 0 \qquad \text{for } k \text{ even;}$$

(ii) if y is an even function of u (even symmetry), then reversing the sign of u leaves y unaltered, whence

$$a_k = b_k = 0 \qquad \text{for } k \text{ odd;}$$

(iii) if y is a single-valued function of u (memoryless nonlinearity), then changing t to $(\pi/\omega) - t$ leaves both u and y unchanged, and so

$$\begin{aligned} a_k &= 0 \qquad \text{for } k \text{ odd,} \\ b_k &= 0 \qquad \text{for } k \text{ even.} \end{aligned}$$

In applications of the describing function method, it is necessary that the input and output signals should be treated consistently, so that, if the input is taken to be a pure sinusoid, the Fourier series for the output should be truncated at $k = 1$. However, in the present case, where the input is assumed to be unbiased (no constant term), the approximation will only be consistent if the output is also unbiased, which is ensured only if the nonlinearity has odd symmetry, so that

$$c_0 = \tfrac{1}{2} a_0 = 0.$$

On the other hand, there is no particular need to assume single-valueness for the nonlinearity, since the possibility of its being multivalued is allowed for by the presence of both sine and cosine terms in the output, which we approximate by writing

$$y(t) \simeq a_1 \cos(\omega t) + b_1 \sin(\omega t)$$
$$= Y \sin(\omega t + \phi)$$

where Y and ϕ are given by

$$a_1 = Y \sin\phi, \qquad b_1 = Y \cos\phi.$$

The effect of the nonlinearity on the sinusoidal input is thus to multiply its amplitude by a gain factor

$$\frac{Y}{U} = \frac{\sqrt{(a_1^2 + b_1^2)}}{U}$$

and advance its phase through an angle

$$\phi = \arctan\left(\frac{a_1}{b_1}\right)$$

so that it has an effective transfer function N (the describing function), given by

$$|N| = \frac{Y}{U}, \qquad \angle N = \phi.$$

If the nonlinearity is single-valued, then $a_1 = 0$ and hence $\phi = 0$, so that the describing function is real; in multivalued cases, it becomes complex, with ϕ usually being negative, meaning that the phase is retarded. From the integral expressions for the Fourier coefficients, it follows that, if the input-output relation is given by

$$y = f(u)$$

then the describing function is

$$N = \frac{b_1 + ia_1}{U} = \frac{1}{\pi U}\int_0^{2\pi} f(U \sin\theta)\,(\sin\theta + \mathrm{i}\cos\theta)\,\mathrm{d}\theta$$

which is in general a function of the input amplitude U, though not of ω, which has disappeared from the expression when we set $\theta = \omega t$.

The foregoing development refers to what is sometimes called the single-input describing function (SIDF), meaning that the input function consists of only one term. Since this excludes the consideration of biased signals and unsymmetrical nonlinearities, both of which are to be expected in practice, the most obvious generalisation is to take the input as

$$u(t) = U_0 + U_1 \sin(\omega t)$$

which leads to the so-called dual-input describing function (DIDF). In this case, we approximate the output signal by

$$\begin{aligned} y(t) &\simeq c_0 + a_1 \cos(\omega t) + b_1 \sin(\omega t) \\ &= Y_0 + Y_1 \sin(\omega t + \phi) \end{aligned}$$

where

$$Y_0 = c_0, \qquad Y_1 \sin\phi = a_1, \qquad Y_1 \cos\phi = b_1.$$

We can then define the DIDF to have two components

$$N_0 = \frac{Y_0}{U_0}, \qquad N_1 = \frac{Y_1 \exp(i\phi)}{U_1}$$

which are given by

$$N_0 = \frac{1}{2\pi U_0}\int_0^{2\pi} f(U_0 + U_1 \sin\theta)\, d\theta$$

$$N_1 = \frac{1}{\pi U_1}\int_0^{2\pi} f(U_0 + U_1 \sin\theta)(\sin\theta + i\cos\theta)\, d\theta$$

both of which are functions of U_0 and U_1. Clearly, N_0 is real but N_1 may be complex; in fact, just as for the SIDF, N_1 will also be real if $f(u)$ is single-valued, since we can then write the expression for the imaginary part, by setting $\upsilon = U_1 \sin\theta$, as

$$\frac{1}{\pi U_1^2}\int_0^0 f(U_0 + \upsilon)\, d\upsilon = 0$$

though this will not hold if $f(u)$ is multivalued. Indeed, the same trick can be used to simplify the calculation of describing functions for a general nonlinearity with memory, where, as shown in Fig. 3.1, the output is given by two different expressions $y = f_+(u)$, $y = f_-(u)$, respectively, according as u is increasing or decreasing. Changing the integration range for θ from $(0, 2\pi)$ to $(-\pi/2, 3\pi/2)$, which makes no difference to the integral, we obtain

$$\begin{aligned} \text{Im}\, N_1 &= \frac{1}{\pi U_1^2}\int_{-U_1}^{U_1} \{f_+(U_0+\upsilon) - f_-(U_0+\upsilon)\}\, d\upsilon \\ &= \frac{\pm\Delta}{\pi U_1^2} \end{aligned}$$

where Δ is the area enclosed by the graph of the input–output relation in the (u,y) plane, with the sign taken positive or negative according as the curve is traversed clockwise or anticlockwise, respectively. By similar means, we can also show that the correct expressions for N_0 and Re N_1 are

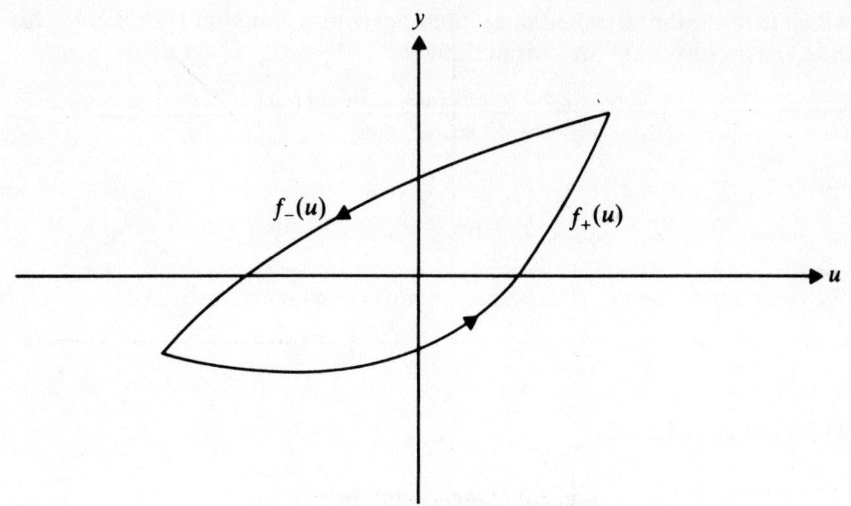

Fig. 3.1 A two-valued nonlinearity.

obtained if $f(u)$ is simply replaced by $\frac{1}{2}\{f_+(u) + f_-(u)\}$, that is to say, the average of the two functions associated with different branches of the input–output characteristic.

It is evident that the above formulation can be generalised further, by retaining more terms in the Fourier series and defining a multiple-input describing function (MIDF). An alternate version of the MIDF is the 'describing function matrix', obtained by truncating an infinite matrix which relates the infinite-dimensional vectors formed from the Fourier coefficients of the input and output signals, respectively. We can also extend the method to deal with cases where the signals contain components at frequencies which are not rationally related, so that they are almost-periodic rather than periodic, although this raises major difficulties in computing the generalised Fourier coefficients of the output, unless the nonlinearity has a simple analytic form. Nevertheless, this approach can still be useful in such applications, provided that one has some prior idea as to the frequencies which are likely to be significant.

Example 3.1 Piecewise-linear functions

Several types of nonlinearity, commonly encountered in control system design, such as dead zones and saturation effects, are often represented by piecewise-linear approximation, so that the input–output characteristic is composed of a number of straight lines joined together. A very simple example of this kind of function is illustrated in Fig. 3.2 and has the mathematical representation

$$y = u\,H(u)$$

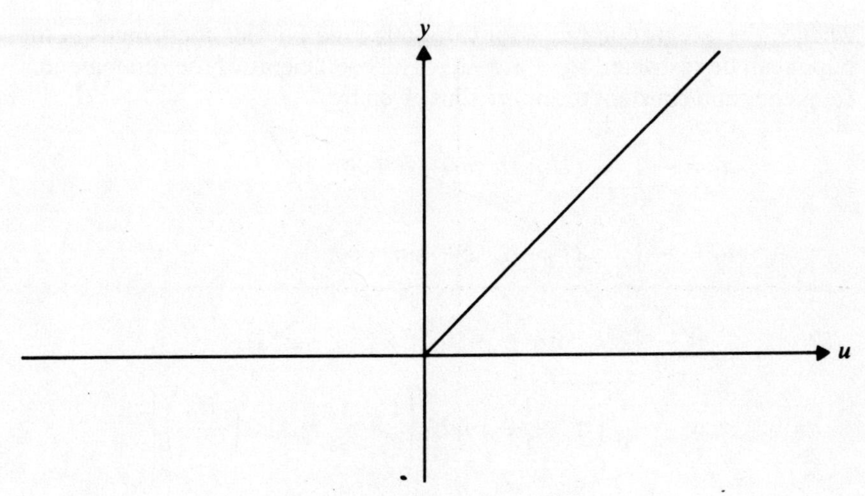

Fig. 3.2 Graph of $y = uH(u)$.

where $H(u)$ denotes the Heaviside unit function

$$H(u) = \frac{1 + \operatorname{sgn}(u)}{2} = \begin{cases} 1 & u > 0, \\ 0 & u < 0, \end{cases}$$

the value of $H(0)$ being irrelevant but usually taken as ½. If a biased sinusoidal input signal

$$u = U_0 + U_1 \sin(\omega t)$$

is applied to this nonlinear element, the ouput will have the form shown in Fig. 3.3 (assuming $U_1 \geqslant |U_0|$), since $u = 0$ when $\omega t = -\alpha$ or $\pi + \alpha$, where

$$\alpha = \arcsin\left(\frac{U_0}{U_1}\right)$$

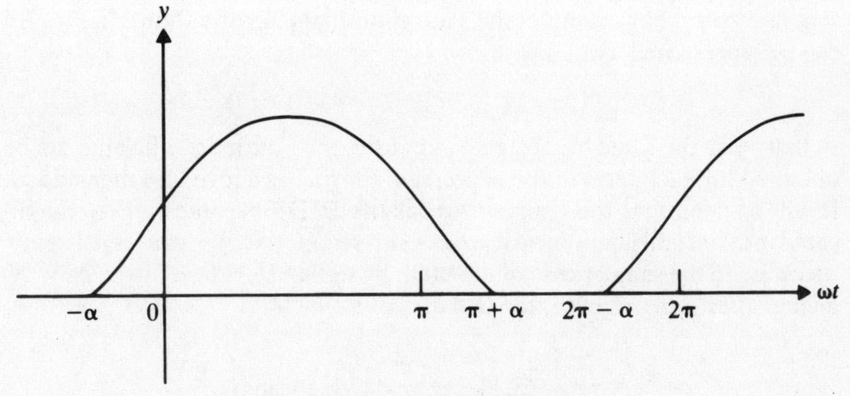

Fig. 3.3 Output waveform for Fig. 3.2.

whence the integration range for calculating the Fourier coefficients of the ouput can be restricted to $(-\alpha, \pi+\alpha)$. The coefficients of the fundamental-frequency and constant terms are thus given by

$$a_1 = \frac{1}{\pi}\int_{-\alpha}^{\pi+\alpha} (U_0+U_1 \sin\theta) \cos\theta \, d\theta = 0,$$

$$b_1 = \frac{1}{\pi}\int_{-\alpha}^{\pi+\alpha} (U_0+U_1 \sin\theta) \sin\theta \, d\theta$$

$$= \frac{1}{\pi}\{2U_0 \cos\alpha + \tfrac{1}{2} U_1 (\pi + 2\alpha - \sin2\alpha)\}$$

$$= \frac{U_0}{\pi}\sqrt{\left(1 - \frac{U_0^2}{U_1^2}\right)} + U_1\left\{\frac{1}{2} + \frac{1}{\pi} \arcsin\left(\frac{U_0}{U_1}\right)\right\},$$

$$c_0 = \frac{1}{2\pi}\int_{-\alpha}^{\pi+\alpha} (U_0+U_1 \sin\theta) \, d\theta$$

$$= \frac{1}{2\pi} \{U_0 (\pi + 2\alpha) + 2U_1 \cos\alpha\}$$

$$= U_0\left\{\frac{1}{2} + \frac{1}{\pi} \arcsin\left(\frac{U_0}{U_1}\right)\right\} + \frac{\sqrt{(U_1^2 - U_0^2)}}{\pi},$$

from which the components of the DIDF are obtained as

$$N_0 = \frac{c_0}{U_0}, \qquad N_1 = \frac{b_1}{U_1}.$$

Since describing functions are linearly related to the nonlinear functions which they describe, it is quite legitimate to decompose complicated nonlinearities into sums of simpler ones and then obtain the complete describing function by adding up the contributions of the separate parts. For example, the saturation characteristic shown in Fig. 3.4 can be represented as

$$f(u) = (u+a) H(u+a) - (u-a) H(u-a) - a$$

so that, with the same input signal as before, the Fourier coefficients can be obtained for each term in the expression for $f(u)$ as above and then added. It will be seen that the expressions for the DIDF become rather complicated but, as this nonlinearity has odd symmetry, we can consistently specialise to the simpler case of the SIDF by setting $U_0 = 0$, $U_1 = U$, $N_1 = N$, giving, after some algebra, for $U \geqslant a$,

$$N = \frac{2}{\pi}\left\{\frac{a}{U}\sqrt{\left(1 - \frac{a^2}{U^2}\right)} + \arcsin\left(\frac{a}{U}\right)\right\}$$

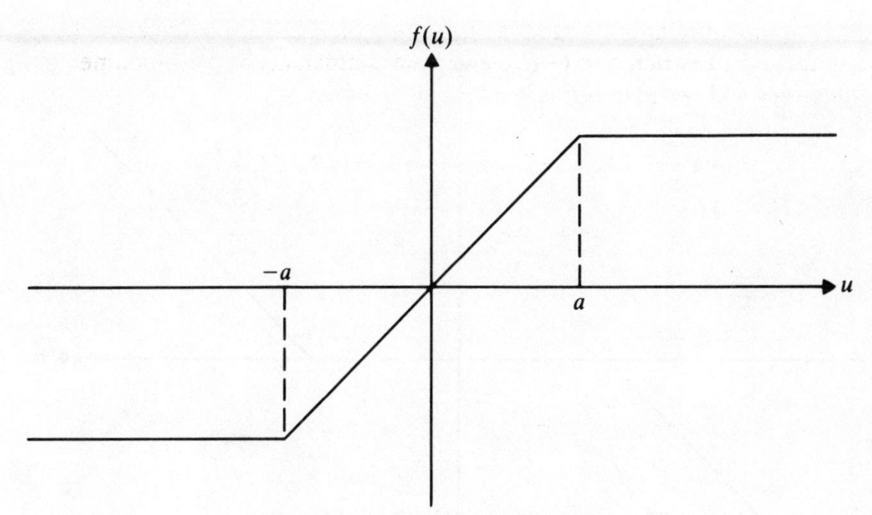

Fig. 3.4 A saturating nonlinearity.

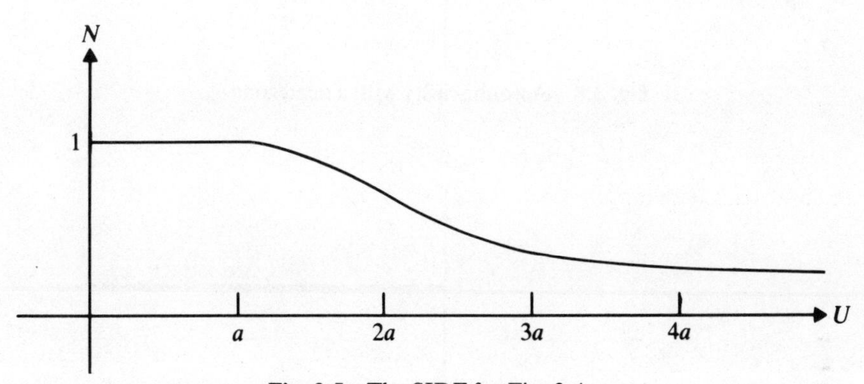

Fig. 3.5 The SIDF for Fig. 3.4.

which is plotted in Fig. 3.5. Of course, if $U < a$, the nonlinearity does not saturate and thus behaves as a linear element with unit gain, giving $N=1$. Similarly, the describing function for the dead zone characteristic in Fig. 3.6 can be calculated easily by noticing that, if the output functions shown in Figs 3.4 and 3.6 are denoted by y_S, y_D, respectively, then $y_D = u - y_S$. Consequently, any describing function component for y_D may be obtained by subtracting the corresponding one for y_S from unity.

Example 3.2 Relay-type nonlinearities

These functions constitute a special case of the piecewise-linear class, being in fact piecewise constant, but discontinuous at certain points. The simplest example is perhaps the Heaviside unit function

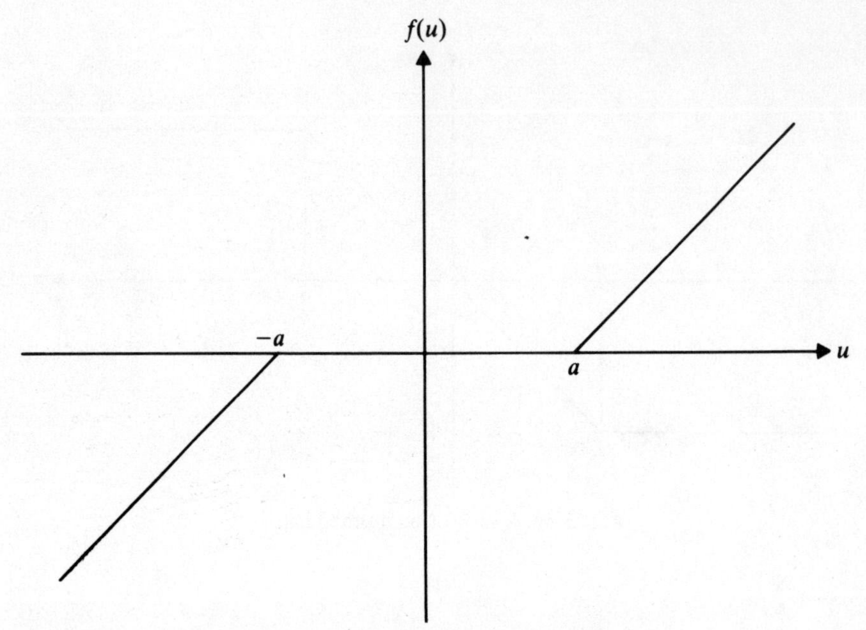

Fig. 3.6 A nonlinearity with a dead zone.

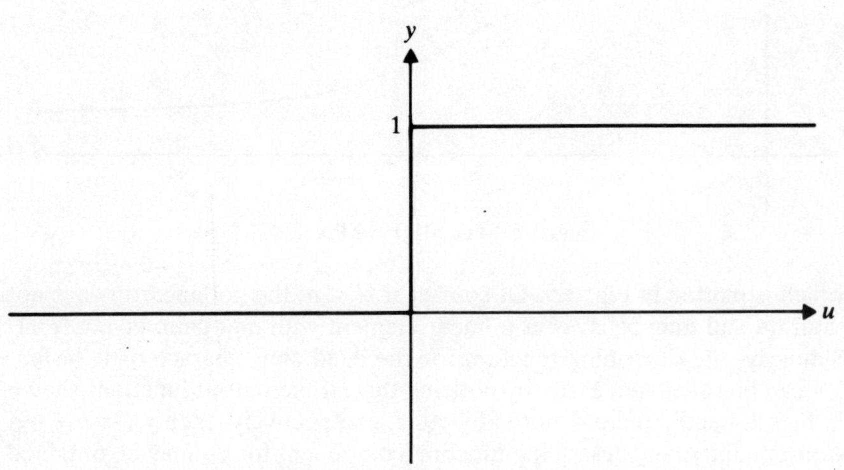

Fig. 3.7 Graph of $y = H(u)$.

$$y = H(u)$$

illustrated in Fig. 3.7. With the same notation for the biased sinusoidal input as in the previous example, we immediately obtain the DIDF components

$$N_0 = \frac{1}{2\pi U_0}\int_{-\alpha}^{\pi+\alpha} d\theta = \frac{\pi+2\alpha}{2\pi U_0} = \frac{1}{U_0}\left\{\frac{1}{2} + \frac{1}{\pi}\arcsin\left(\frac{U_0}{U_1}\right)\right\},$$

$$N_1 = \frac{1}{\pi U_1}\int_{-\alpha}^{\pi+\alpha} \sin\theta \, d\theta = \frac{2\cos\alpha}{\pi U_1} = \frac{2}{\pi U_1}\sqrt{\left(1 - \frac{U_0^2}{U_1^2}\right)},$$

for $U_1 \geqslant |U_0|$, the output signal being a sequence of square pulses as in Fig. 3.8.

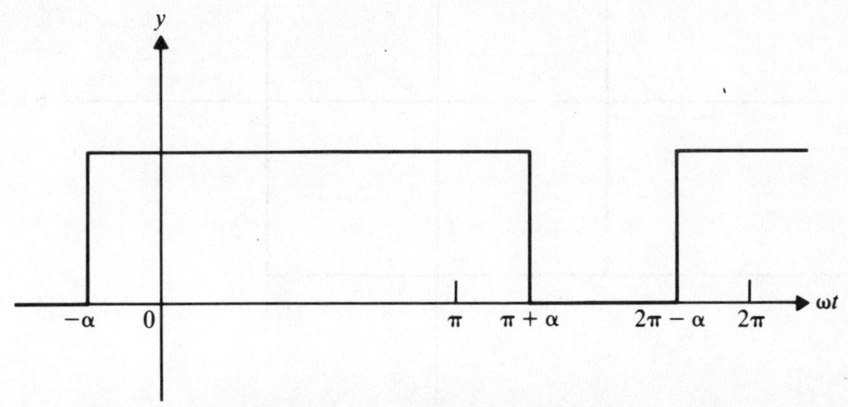

Fig. 3.8 Output waveform for Fig. 3.7.

A more complicated example is provided by the two-valued characteristic shown in Fig. 3.9, which has an overlap region, giving rise to hysteretic behaviour if the range of the input signal covers the interval $(-a,a)$, that is to say, if $U_1 \geqslant a + |U_0|$. In this case, the real components of the DIDF can be computed from the mean of the two output functions, which is given by

$$y_M = H(u-a) + H(u+a) - 1$$

as represented in Fig. 3.10. We thus obtain

$$N_0 = \frac{1}{\pi U_0}\left\{\arcsin\left(\frac{U_0-a}{U_1}\right) + \arcsin\left(\frac{U_0+a}{U_1}\right)\right\},$$

$$\text{Re}\, N_1 = \frac{2}{\pi U_1}\left\{\sqrt{\left(1 - \frac{(U_0-a)^2}{U_1^2}\right)} + \sqrt{\left(1 - \frac{(U_0+a)^2}{U_1^2}\right)}\right\}$$

but there is also an imaginary part, arising from the hysteresis loop, which encloses an area $\Delta = 4a$, encircled anticlockwise, whence

$$\text{Im}\, N_1 = \frac{-4a}{\pi U_1^2}.$$

Since the input–output relation has odd symmetry, we can also immediately obtain the SIDF, which is given, for $U \geqslant a$, by

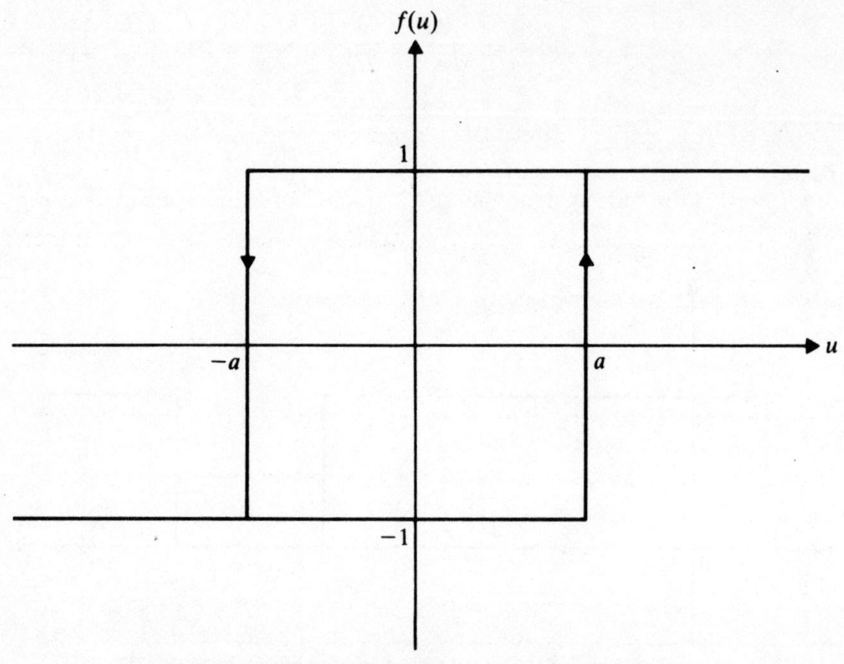

Fig. 3.9 A relay characteristic with an overlap region.

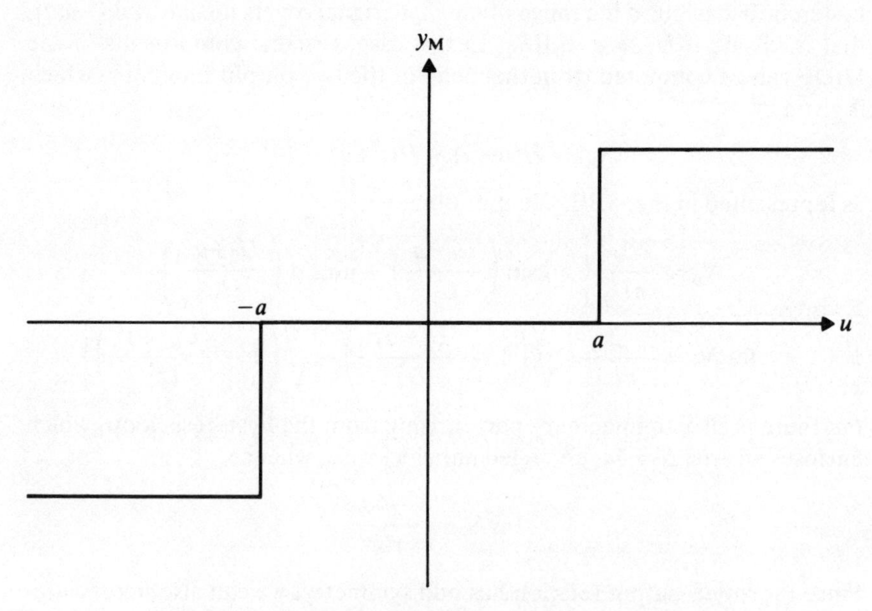

Fig. 3.10 Graph of mean output function for Fig. 3.9.

$$N = \frac{4}{\pi U}\left\{\sqrt{\left(1 - \frac{a^2}{U^2}\right)} - \frac{ia}{U}\right\}$$

as illustrated in Fig. 3.11.

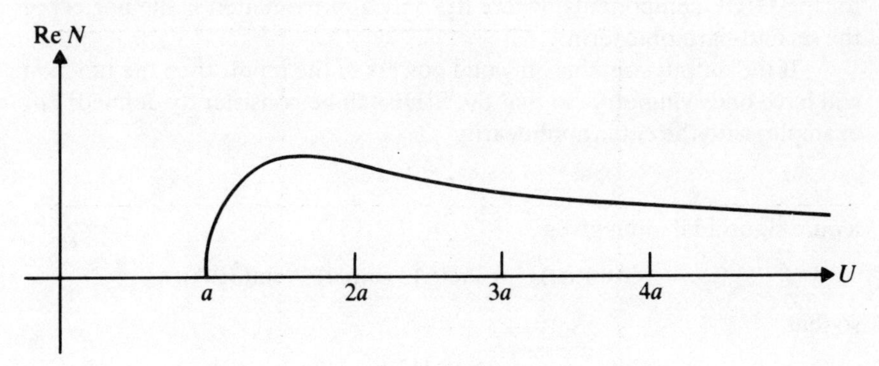

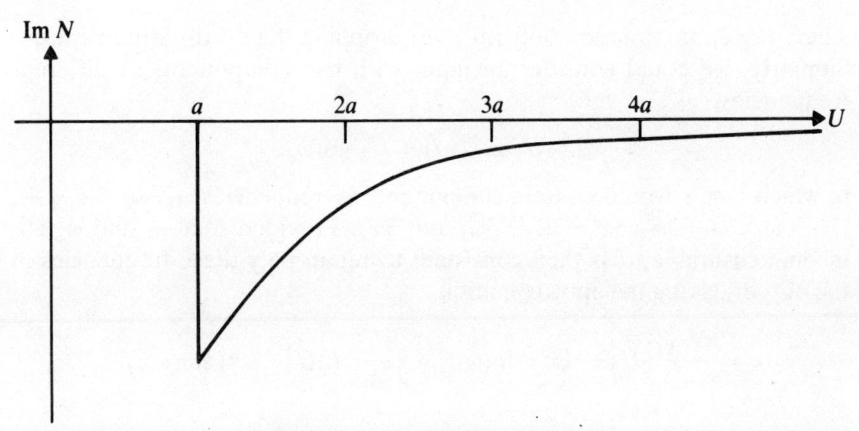

Fig. 3.11 The SIDF for Fig. 3.9.

Example 3.3 Polynomials

In cases where the output of a nonlinear element can be expressed as a polynomial in the input, it may be simpler to calculate the describing function by trigonometric means, rather than integration. Thus, for the quadratic nonlinearity

$$y = u^2$$

the input of a biased sinusoid yields

$$\begin{aligned} y &= \{U_0 + U_1 \sin(\omega t)\}^2 \\ &= U_0^2 + \tfrac{1}{2} U_1^2 + 2U_0U_1 \sin(\omega t) - \tfrac{1}{2} U_1^2 \cos(2\omega t) \end{aligned}$$

giving

$$N_0 = U_0 + \frac{U_1^2}{2U_0}, \qquad N_1 = 2U_0,$$

for the DIDF components, where the only approximation is the neglect of the second-harmonic term.

If the output contains only odd powers of the input, then the function will have odd symmetry, so that the SIDF can be consistently defined. For example, with the cubic nonlinearity

$$y = u^3$$

a pure sinusoidal input gives

$$y = \{U\sin(\omega t)\}^3 = \tfrac{1}{4}U^3\,\{3\sin(\omega t) - \sin(3\omega t)\}$$

so that

$$N = 3U^2/4$$

where the approximation only involves dropping the third-harmonic term. Similarly, we could consider an input with two components of different frequencies,

$$u = U_1\sin(\omega_1 t) + U_2\sin(\omega_2 t)$$

in which case y would contain components at frequencies ω_1, ω_2, $2\omega_1+\omega_2$, $|2\omega_1-\omega_2|$, $\omega_1+2\omega_2$, $|\omega_1-2\omega_2|$, $3\omega_1$ and $3\omega_2$. Provided that ω_1 and ω_2 are incommensurable, it is then consistent to retain only these frequencies in the output, giving the approximation

$$y \simeq \frac{3U_1}{4}(U_1^2+2U_2^2)\sin(\omega_1 t) + \frac{3U_2}{4}(2U_1^2+U_2^2)\sin(\omega_2 t).$$

On the other hand, if $\omega_2 = 3\omega_1$ for instance, then further terms must be retained, and also phase relations considered by allowing for cosine terms in u and y.

3.2 Oscillations in feedback systems

One of the principal applications of the describing function method is to the investigation of limit cycles in nonlinear feedback systems. The basic configuration, in the simplest case, is as shown in Fig. 3.12, where there is a single nonlinearity whose output $f(u)$ is fed into a linear element with transfer function $G(s)$, generating a signal y which is then subtracted from an external reference input r, giving $u = r-y$ as the input to the nonlinear element. For the present, we will take the reference signal to be constant,

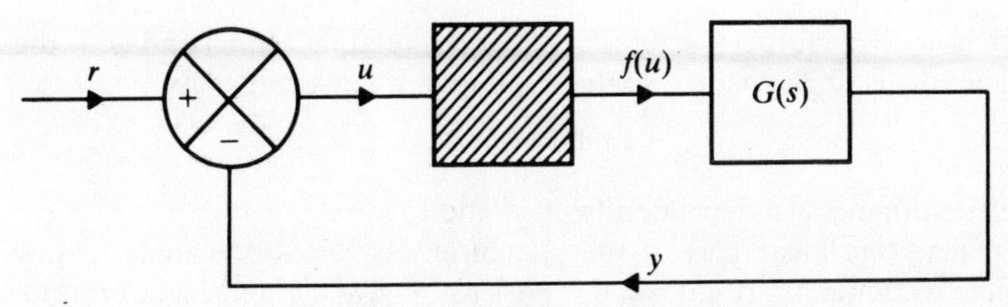

Fig. 3.12 A nonlinear feedback system.

so that the system is effectively autonomous, although the method can be extended to deal with forced systems as well. We assume that, if a limit cycle occurs, it can be adequately approximated by a sinusoidal oscillation, so that we can take

$$u \simeq U_0 + U_1 \sin(\omega t)$$

and correspondingly

$$f(u) \simeq N_0 U_0 + (\mathrm{Re}\, N_1)\, U_1 \sin(\omega t) + (\mathrm{Im}\, N_1)\, U_1 \cos(\omega t)$$

where N_0 and N_1 are the DIDF components. In complex notation, this becomes

$$f(u) \simeq N_0 U_0 + \mathrm{Im}\{N_1 \exp(\mathrm{i}\omega t)\}\, U_1$$

and hence

$$y \simeq G(0)\, N_0 U_0 + \mathrm{Im}\{G(\mathrm{i}\omega)\, N_1 \exp(\mathrm{i}\omega t)\}\, U_1$$

so that, setting

$$u + y = r$$

and equating coefficients of corresponding terms, we get

$$\begin{aligned} \{1 + G(0)\, N_0\}\, U_0 &= r \\ 1 + \mathrm{Re}\{G(\mathrm{i}\omega)\, N_1\} &= 0 \\ \mathrm{Im}\{G(\mathrm{i}\omega)\, N_1\} &= 0 \end{aligned}$$

where we have assumed $U_1 \neq 0$ in order that the oscillatory part of the solution should not vanish. Displaying the dependence of the describing function components on U_0 and U_1 explicitly, we can rewrite the first of these equations as

$$G(0) = \frac{1}{N_0(U_0, U_1)} \left(\frac{r}{U_0} - 1 \right)$$

giving a functional relation between U_0 and U_1, while the other two equations can be combined to give

$$G(\mathrm{i}\omega) = \frac{-1}{N_1(U_0, U_1)}$$

which is equivalent to a pair of real equations relating U_0, U_1 and ω. If the nonlinearity is single-valued, so that N_1 is real, we evidently have

$$\operatorname{Im} G(i\omega) = 0$$

which determines ω independently of U_0 and U_1.

Since the linear part of the system enters the above analysis only through its frequency response, it is possible to give a graphical interpretation of the limit cycle conditions by using the Nyquist diagram, that is to say, the plot of $G(i\omega)$ in the complex plane as ω goes from 0 to ∞. For this purpose, we suppose that the equation involving $G(0)$ has been solved for U_0 as a function of U_1, giving

$$U_0 = F(U_1)$$

and we then define

$$N(U_1) \equiv N_1(F(U_1), U_1)$$

so that the other equation for the limit cycle becomes

$$G(i\omega) = \frac{-1}{N(U_1)}$$

where the definition of N is such that it coincides with the SIDF in the case of a nonlinearity with odd symmetry and zero reference input. The existence of a limit cycle is thus predicted if there is an intersection between the loci, in the complex plane, of $G(i\omega)$ and $-1/N(U_1)$, which can be independently plotted since they depend on different variables. Moreover, estimates for the parameters of the limit cycle can be obtained from the values taken by the independent variables at the point of intersection, ω being the angular frequency and U_1 the amplitude of the oscillation. Further, it is also possible to make a prediction as to whether or not the limit cycle will be stable, as follows. We assume that, if the system is slightly disturbed from its state of steady oscillation, the nonlinear part can still be adequately represented by its describing function, so that the parameters of the perturbed motion are related by

$$G(i\omega + \Delta s) = \frac{-1}{N(U_1 + \Delta U)}$$

where ΔU is the change in amplitude, $\operatorname{Im}\Delta s$ is the frequency deviation and $\operatorname{Re}\Delta s$ is the growth rate. Hence, for small perturbations, we have

$$\begin{aligned} \frac{\partial G}{\partial s}(i\omega)\,\Delta s &= \frac{\partial}{\partial U_1}\left(\frac{-1}{N(U_1)}\right)\Delta U \\ &= -i\,\frac{\partial G(i\omega)}{\partial \omega}\,\Delta s \end{aligned}$$

giving

$$\frac{\Delta s}{\Delta U} = \mathrm{i}\frac{\partial}{\partial U_1}\left(\frac{-1}{N(U_1)}\right)\left(\frac{\partial G(\mathrm{i}\omega)}{\partial \omega}\right)^{-1}$$

and, for the limit cycle to be stable, we require $\mathrm{Re}\Delta s$ and ΔU to have opposite signs, so that a positive or negative amplitude deviation causes the oscillation to decay or grow, respectively, back to its steady condition. Now, defining the complex quantities

$$\xi = \frac{\partial G(\mathrm{i}\omega)}{\partial \omega}, \qquad \eta = \frac{\partial}{\partial U_1}\left(\frac{-1}{N(U_1)}\right),$$

which point, repsectively, in the directions along the loci of $G(\mathrm{i}\omega)$ for increasing ω and $-1/N(U_1)$ for increasing U_1, it follows that

$$\frac{\mathrm{Re}\Delta s}{\Delta U} = -\mathrm{Im}\left(\frac{\eta}{\xi}\right)$$

so the necessary condition for limit cycle stability becomes $\mathrm{Im}(\eta/\xi) > 0$ or, equivalently,

$$0 < \underline{/\eta} - \underline{/\xi} < \pi.$$

This result has an immediate graphical interpretation in terms of the way in which the loci intersect, as illustrated in Fig. 3.13: the limit cycle is predicted to be stable or unstable according as the locus of $-1/N$ crosses the locus of G (the Nyquist plot) from right to left or from left to right, respectively, as U_1 increases, viewed along the direction of increasing ω. Although we have only considered in detail the case of a system with a single nonlinearity, using the SIDF or DIDF approximation, the method can readily be generalised to cover both multiple nonlinearities and the inclusion of other frequencies, though a simple graphical analysis is not usually possible in such cases.

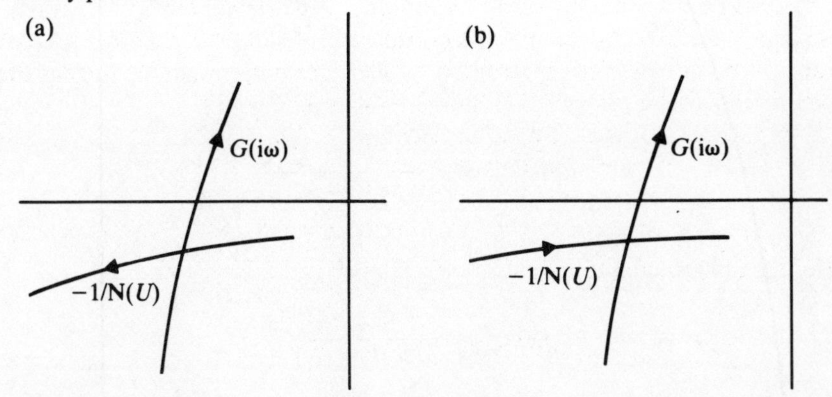

Fig. 3.13(a) Prediction of a stable limit cycle; (b) prediction of an unstable limit cycle.

Example 3.4 Saturation in a feedback loop

Consider the feedback system of Fig. 3.12, where the nonlinearity is of the saturating type, with the characteristic shown in Fig. 3.4, and the linear element has the transfer function

$$G(s) = \frac{K}{s(s^2+bs+c)}$$

where K, b and c are positive constants. Since $G(s)$ becomes infinite as $s \to 0$, it follows from the limit cycle equations, in the DIDF approximation, that $U_0 = 0$ for any constant reference input r, and hence the condition for an oscillation to exist is that the loci of $G(i\omega)$ and $-1/N(U_1)$ should intersect, where N is the SIDF shown in Fig. 3.5. The loci are illustrated in Fig. 3.14, with $G(i\omega)$ crossing the negative real axis at $-K/bc$, when $\omega = \sqrt{c}$, so that an intersection occurs if

$$K > bc$$

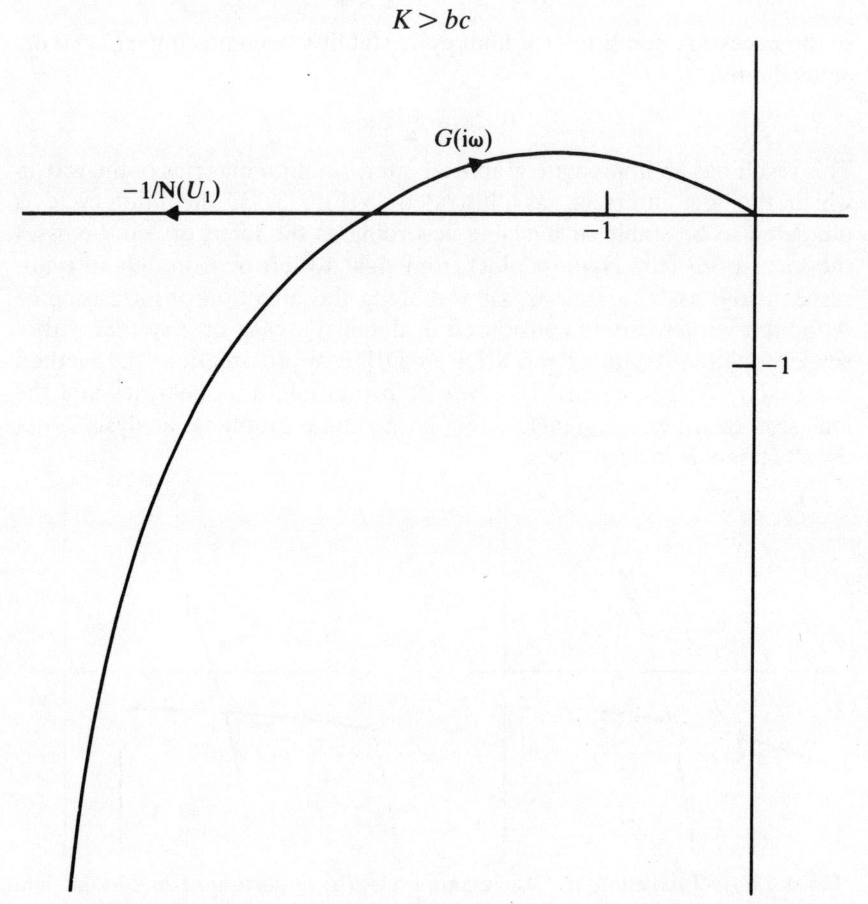

Fig. 3.14 Nyquist diagram for Example 3.4 ($b = 2$, $c = 2$, $K = 10$).

which is just the Routh–Hurwitz condition for the equilibrium state to be unstable. It is thus predicted that a stable limit cycle will appear whenever the gain K is high enough to destabilise the equilibrium condition.

Example 3.5 Feedback system containing a relay

Again we take the configuration of Fig. 3.12, with the nonlinear element now being a relay device, having the characteristic of Fig. 3.9, and the transfer function $G(s)$ being given by

$$G(s) = \frac{K}{s(1+s)}$$

where K is a positive constant. As in the previous example, $G(s) \to \infty$ as $s \to 0$, so that we can again set $U_0 = 0$ and use the SIDF approximation for limit cycle prediction. In this case, N is the complex function given at the end of Example 3.2, whence

$$\frac{1}{N(U_1)} = \frac{\pi}{4}\{\sqrt{(U_1^2 - a^2)} + ia\}$$

for $U_1 \geqslant a$. The loci of $G(i\omega)$ and $-1/N(U_1)$ are as shown in Fig. 3.15, from which we see that an intersection always exists, with the orientation of the

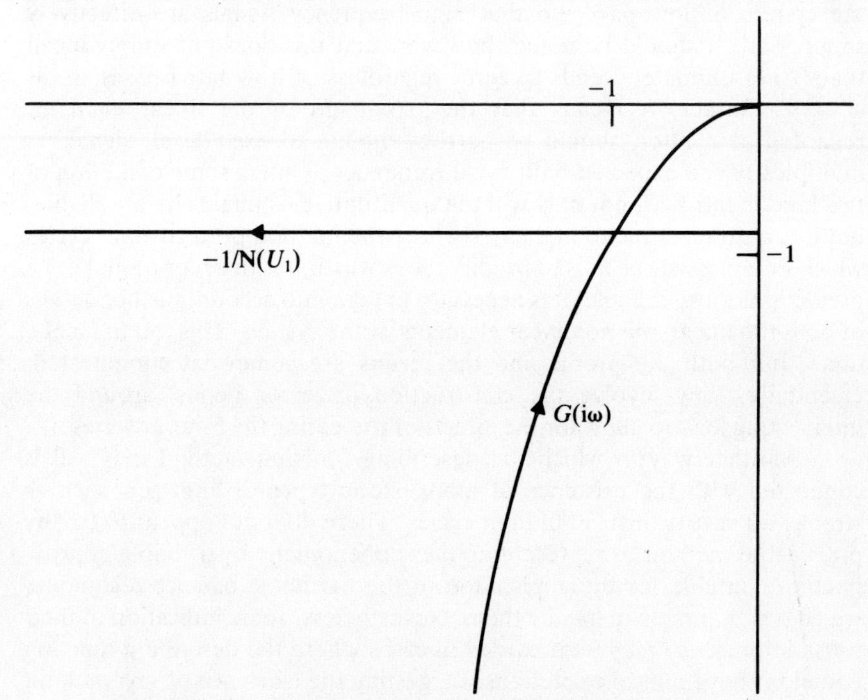

Fig. 3.15 Nyquist diagram for Example 3.5 ($K = 2$, $a = 1$).

plots being such that a stable limit cycle is predicted. To estimate its parameters, we use the real and imaginary parts of the equation $G = -1/N$, namely

$$\frac{K}{1+\omega^2} = \frac{\pi}{4}\sqrt{(U_1^2 - a^2)}$$

$$\frac{K}{\omega(1+\omega^2)} = \frac{\pi a}{4},$$

from which ω is found by solving the second equation, and U_1 by substitution into the first one.

3.3 Validity of the describing function approximation

Although the describing function method is intuitively appealing and convenient for use, it is important to bear in mind the limitations which arise from its approximate nature. The method depends crucially on the irrelevance of the neglected higher harmonics, and can thus be expected to perform best when the nonlinearity is smooth and the linear component of the system is 'low pass', so that high frequency signals are effectively suppressed. It should be noted, however, that this does not simply mean that $G(i\omega)$ ultimately tends to zero, regardless of how fast it does so, as $\omega \to \infty$. Rather, it means that the pass-band of the linear element, regarded as a filter, should be narrow enough to exclude all signals at multiples of the expected limit cycle frequency. Unless some condition of this kind is satisfied, not only will the quantitative estimates be unreliable, but it is quite possible to find cases where the method predicts limit cycles which do not exist, or fails to predict those which do. In order to make the predictions more reliable, it is necessary to take into account further details of both the linear and nonlinear elements of the system. This can in fact be done, but both the proofs and the results are somewhat complicated; essentially, they involve the construction of 'error bands' around the intersecting loci, to allow for the effects of truncating the Fourier series.

Another way in which the describing function method may fail is connected with the existence of more exotic types of limit set, such as strange attractors, instead of limit cycles. There does not appear to be any practicable method of representing these phenomena by a simple approximation, suitable for the application of the harmonic balance technique, which is thus unable to handle them. Nevertheless, some indication of their possible presence may be provided in cases where the describing function equations have multiple solutions, suggesting the existence of several limit cycles. If these are predicted to lie close together, it may well turn out that they do not actually exist as such, but are replaced by some form of chaotic behaviour.

In view of the fact that all predictions based on harmonic balancing are to some extent unreliable, it is to be expected that those concerning the stability of the solution will be even more so, since they involve not only the truncation of the Fourier series, but also the assumption that the describing function method is still valid even when applied to signals whose amplitudes are exponentially growing or decaying. The adequacy of this concept is very difficult to assess, in general, and it is probably better to regard such predictions as giving no more than a rough indication of what to expect. Often, in fact, it may be clear on other grounds whether or not an oscillation is likely to be stable, and if this expectation is reinforced by the graphical prediction, then it should be reliable; otherwise, a deeper analysis is called for. This, however, is usually no easy matter, since it means resorting to the underlying differential equation system, which is nonlinear and could be of high order.

Example 3.6 Van der Pol's equation

One form of the Van der Pol equation for a self-exciting oscillator, studied in Example 2.2, namely

$$\ddot{y} + \epsilon(3y^2 - 1)\dot{y} + y = 0$$

can be associated directly with the feedback system of Fig. 3.12, by taking the transfer function as

$$G(s) = \frac{\epsilon s}{s^2 - \epsilon s + 1}$$

with the nonlinearity

$$f(u) = u^3$$

and zero reference input, so that $u = -y$. From Example 3.3, the SIDF is $N = 3U^2/4$, and the loci of $G(i\omega)$ and $-1/N$ are as shown in Fig. 3.16, so that a stable oscillation is predicted with angular frequency $\omega = 1$ (period 2π) and amplitude $U = 2/\sqrt{3}$.

Now, although Van der Pol's equation is not analytically soluble, it is possible to obtain asymptotically exact expressions for the limit cycle parameters as ϵ approaches zero or infinity. In the small-parameter limit ($\epsilon \to 0$), the equation becomes that of a simple harmonic oscillator with unit angular frequency, coinciding with the prediction of the describing function method. To analyse the large-parameter case ($\epsilon \to \infty$), we set up a state-space representation as in Example 2.2, and rewrite the equations as

$$\frac{dv}{d\tau} = y$$

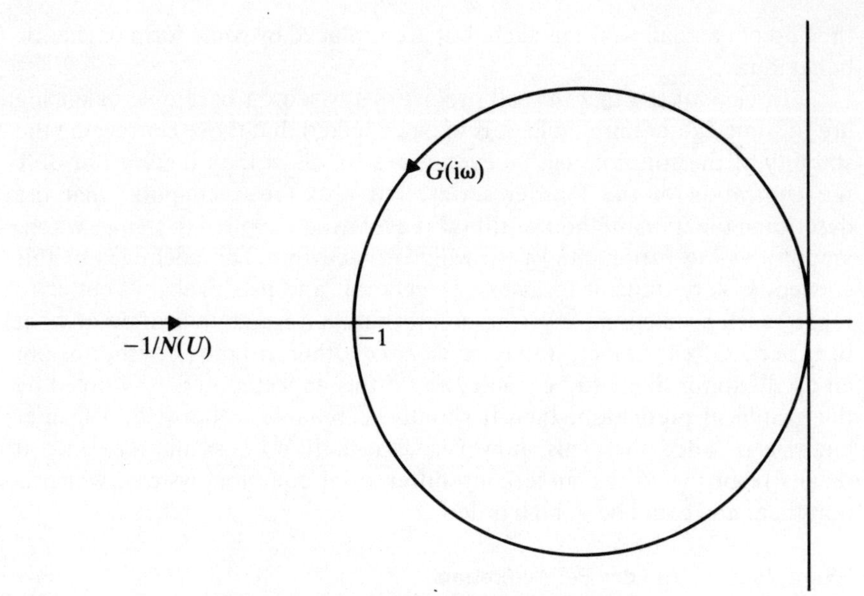

Fig. 3.16 Nyquist diagram for Example 3.6.

$$\frac{dy}{d\tau} = \epsilon^2\{-v + (y - y^3)\}$$

where $\tau = t/\epsilon$, $v = x_1/\epsilon$, and $y = x_2$ as before. The result, in the limit, is that $dy/d\tau$ can only remain finite when $v = y - y^3$, so that the trajectory consists of segments of this curve, as illustrated in Fig. 3.17, joined by vertical sections which are traversed instantaneously (on the timescale associated with τ). On the curve segments, we have

$$y = \frac{dv}{d\tau} = (1 - 3y^2)\,\frac{dy}{d\tau}$$

whence the duration (in τ) of a half-cycle is

$$\int_{2/\sqrt{3}}^{1/\sqrt{3}} \left(\frac{1}{y} - 3y\right) dy = \frac{3}{2} - \ln 2$$

and so the complete period (in t) becomes

$$(3 - 2\ln 2)\epsilon \simeq 1.614\epsilon.$$

Consequently, when ϵ is very large, the describing function prediction for the period becomes wildly inaccurate, since it remains fixed at $2\pi \simeq 6.283$, independent of ϵ. In order to understand the reason for this, let us take a closer look at the 'frequency response' of the linear component of the system, which can be written as

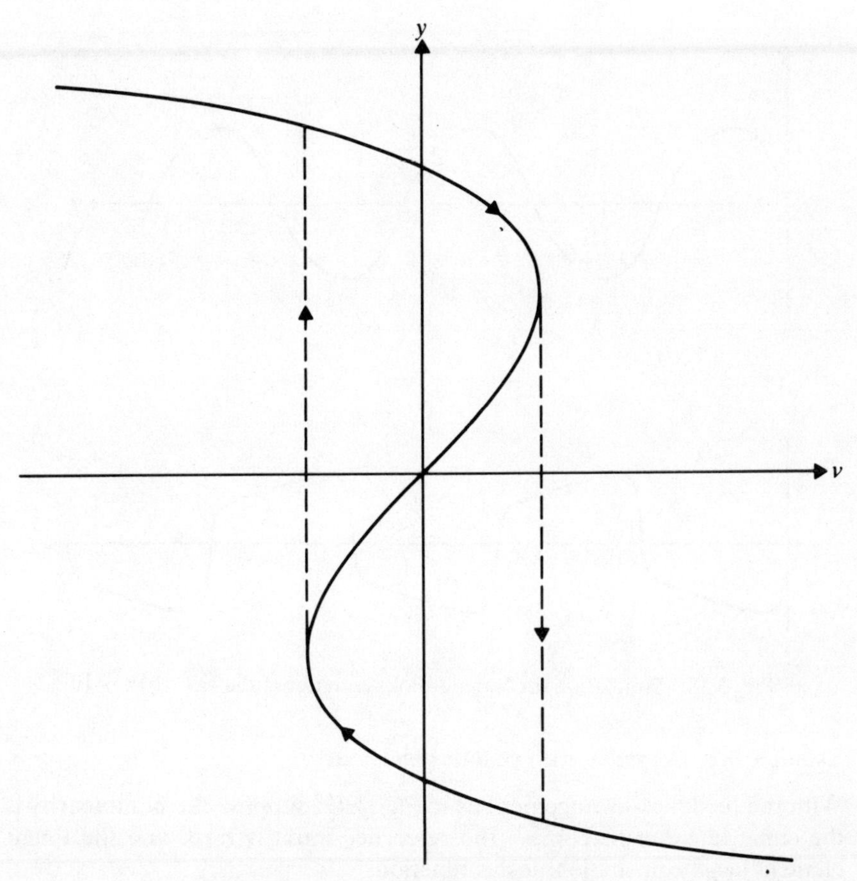

Fig. 3.17 Phase-plane diagram for Van der Pol system in the large-parameter limit.

$$G(i\omega) = \left\{-1 + \frac{i}{\epsilon}\left(\omega - \frac{1}{\omega}\right)\right\}^{-1}$$

ignoring the fact that this element is unstable, since it is irrelevant to our present purpose. It is clear that, as ϵ increases, so does the range of ω over which $G(i\omega) \simeq -1$, to the extent that, in the limit of infinite ϵ, we obtain an 'all-pass' filter, and hence the harmonic content of the limit cycle may become such that the describing function approximation cannot be expected to be valid. This is borne out by a comparison of the shapes of the oscillations for large and small values of ϵ, as shown in Fig. 3.18, even though, somewhat surprisingly, the amplitude prediction remains approximately correct throughout. From the calculation of the limit cycle period in the large-parameter limit, it follows that the fundamental frequency ω_0 varies as $\omega_0 \simeq 3.9/\epsilon$ for $\epsilon \to \infty$, and thus stays well within the pass-band, since $|G(i\omega_0)| \simeq 0.97$.

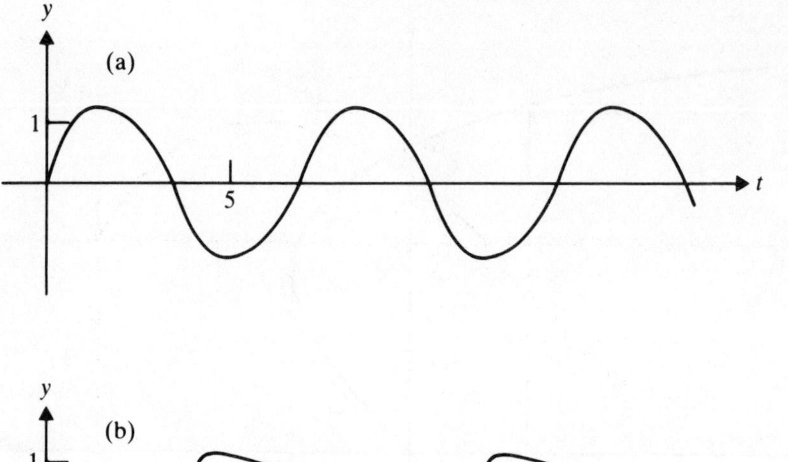

Fig. 3.18 Simulation for Van der Pol's equation: (a) $\epsilon = 1$; (b) $\epsilon = 10$.

Example 3.7 A system with chaotic behaviour

With the feedback arrangement as in Fig. 3.12, suppose the nonlinearity is the cubic function $f(u) = u^3$, the reference input is zero, and the linear element has the unstable transfer function

$$G(s) = \frac{K}{s^3+s^2+cs-K}$$

where c and K are positive constants, so that $G(0) = -1$, and the Nyquist plot, shown in Fig. 3.19, crosses the real axis at $G(\mathrm{i}\sqrt{c}) = -K/(c+K)$. It thus intersects the plot of $-1/N(U)$, where $N(U)$ is the SIDF, when $U = 2\sqrt{\{(c+K)/3K\}}$ and, from the way in which the loci cross, an unstable limit cycle is predicted. However, although the nonlinearity has odd symmetry, it may still be worthwhile to investigate the possibility of other (biased) sinusoidal solutions occurring as well, using the DIDF, which is obtained from the approximation

$$\{U_0+U_1\sin(\omega t)\}^3 \simeq (U_0^3 + \tfrac{3}{2}\,U_0U_1^2) + (3U_0^2U_1 + \tfrac{3}{4}\,U_1^3)\sin(\omega t)$$

where the second and third harmonics have been dropped. This gives

$$N_0 = U_0^2 + \tfrac{3}{2}\,U_1^2, \qquad N_1 = 3U_0^2 + \tfrac{3}{4}\,U_1^2,$$

to be substituted into the limit cycle equations

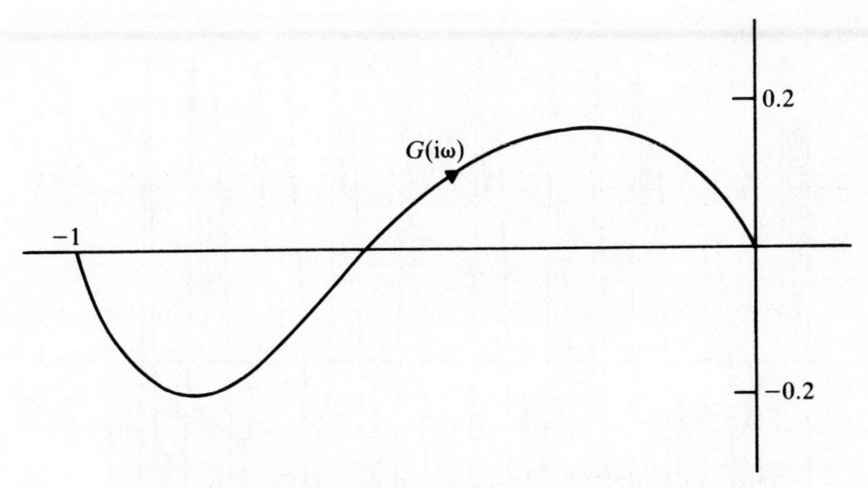

Fig. 3.19 Nyquist plot for Example 3.7 ($c = 1.25$, $K = 1.8$).

$$G(0)N_0 = -1, \qquad G(\mathrm{i}\surd c)\, N_1 = -1,$$

the first of which yields

$$U_0^2 = 1 - 3/2\, U_1^2$$

whence the second becomes

$$G(\mathrm{i}\surd c) N(U_1) = -1$$

where now

$$N(U_1) = 3 - 15/4\, U_1^2$$

so that the loci of G and $-1/N$ intersect when $U_1 = 2\surd\{(2K-c)/15K\}$, provided that $2K>c$, which is the condition for the nonzero equilibrium points ($u = \pm 1$) to be unstable. From the orientation of the intersecting loci, it is predicted that two stable limit cycles then exist, having $U_0 = \pm\surd\{(K+2c)/5K\}$, whose appearance is associated with the onset of instability at the equilibria, a process sometimes known as 'bifurcation'. Simulation studies confirm this impression for certain parameter values, when the equilibria are only slightly unstable and the limit cycles lie close to them. On the other hand, as c/K is made smaller, the oscillations increase in amplitude to the extent that they appear to 'interfere' with each other and merge into a chaotic regime, as illustrated in Fig. 3.20.

3.4 Forced systems

Applications of the harmonic balance method are not restricted to autonomous systems, as described above, but can be made also in cases where

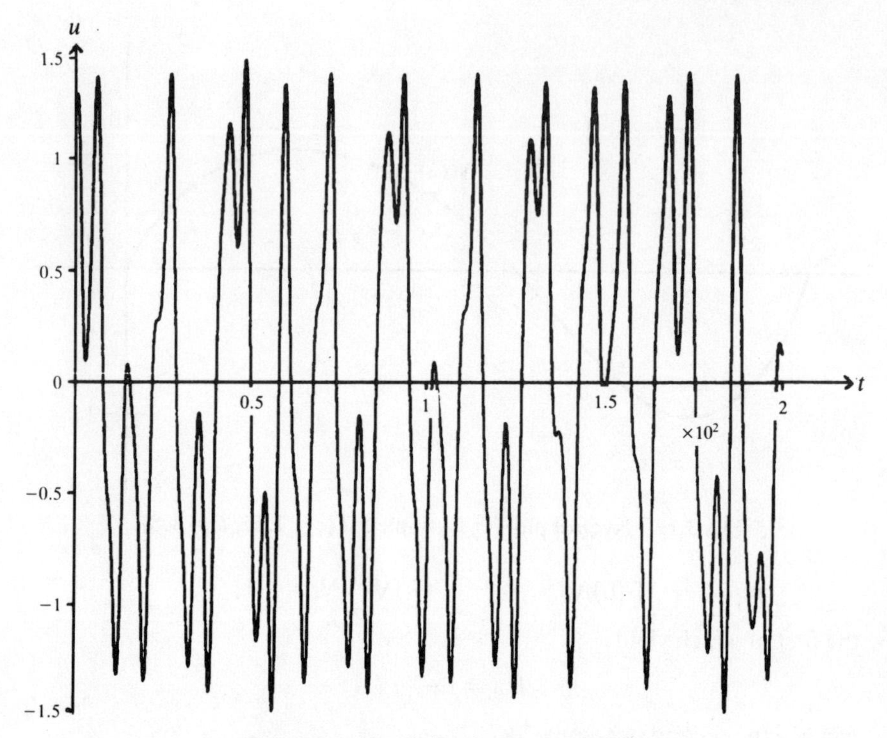

Fig. 3.20 Simulation for Example 3.7.

there is an external input (forcing function), provided that it consists only of a few sinusoidal components. We shall consider here only the case of a simple biased sinusoid,

$$r(t) = R_0 + R_1 \sin(\omega t)$$

inserted as reference input into the system of Fig. 3.12. This, however, is already sufficient to generate several different nonlinear phenomena, depending on the detailed dynamics of the system. The most obvious possibility is that all signals in the feedback loop can be approximated by biased sinusoids of the same frequency, so that

$$\begin{aligned} u(t) &\simeq U_0 + U_1 \sin(\omega t + \psi) \\ &= U_0 + U_1 \operatorname{Im}\{\exp \mathrm{i}(\omega t + \psi)\} \end{aligned}$$

where the angle ψ is introduced to allow for a possible phase difference between $u(t)$ and $r(t)$. Using the DIDF approximation, we then get

$$y(t) \simeq G(0)N_0U_0 + \operatorname{Im}\{G(\mathrm{i}\omega)N_1 \exp \mathrm{i}(\omega t + \psi)\}U_1$$

whence, by equating coefficients,

$$\begin{aligned} \{1 + G(0)N_0\}U_0 &= R_0 \\ \{1 + G(\mathrm{i}\omega)\, N_1\}U_1 \exp(\mathrm{i}\psi) &= R_1 \end{aligned}$$

giving a set of three independent real equations to be solved for U_0, U_1, ψ, when (R_0, R_1, ω) are specified. It should be noted that, in contrast with the limit cycle conditions studied previously, the frequency is now a given parameter, associated with a signal which is externally imposed and not generated within the system. When the equations have been solved, a kind of generalised frequency response may be defined for the closed-loop system, by expressing the quantity $U_1 \exp(i\psi)/R_1$ as a function of ω, with R_0 and R_1 fixed. Unlike the transfer function for a linear system, however, it not only depends on the amplitude and bias of the input signal, but also may be a multivalued function of the frequency. If this happens, it can bring about the so-called 'jump resonance' phenomenon, in which the frequency response has more than one branch and switches from one to another as ω is varied.

As in the case of the limit cycle conditions, a graphical analysis can help to clarify the properties of the forced system solutions. If we again use $N(U_1)$ to denote the value of $N_1(U_0, U_1)$ when U_0 is given as a function of U_1 by solving the first equation, then the second equation can be rewritten as

$$\{1 + G(i\omega)\, N(U_1)\}\, U_1 = R_1 \exp(-i\psi)$$

which can be solved by plotting the left-hand side as a complex function of U_1, for any fixed ω, and finding where it intersects the circle of radius R_1, centred at the origin. Further, for any particular intersection, it is clear that the corresponding solution U_1 will be (locally) an increasing or decreasing function of R_1, according as the locus passes out of or into the circle, as U_1 increases. Since the 'normal' situation, by analogy with linear systems, is for U_1 to increase with R_1, it is plausible to expect that solutions with this property will be stable and those with the contrary behaviour unstable, as illustrated in Fig. 3.21.

Since the external input is now a function of time, we should also consider cases where it enters at a different point in the feedback loop, as this may make a nontrivial difference. Accordingly, we rearrange the configuration as shown in Fig. 3.22, so that, in the DIDF approximation, the harmonic balance equations are now

$$\{G^{-1}(0) + N_0\}\, U_0 = R_0$$
$$\{G^{-1}(i\omega) + N_1\}\, U_1 \exp(i\psi) = R_1$$

whence, again defining $N(U_1)$ by solving the first equation and substituting for U_0 into $N_1(U_0, U_1)$, the second equation becomes

$$\{G^{-1}(i\omega) + N(U_1)\}\, U_1 = R_1 \exp(-i\psi)$$

which can be analysed graphically as before. The two configurations are, in fact, essentially equivalent, since a given forcing function entering at the input of the linear element has the same effect as a different one added to

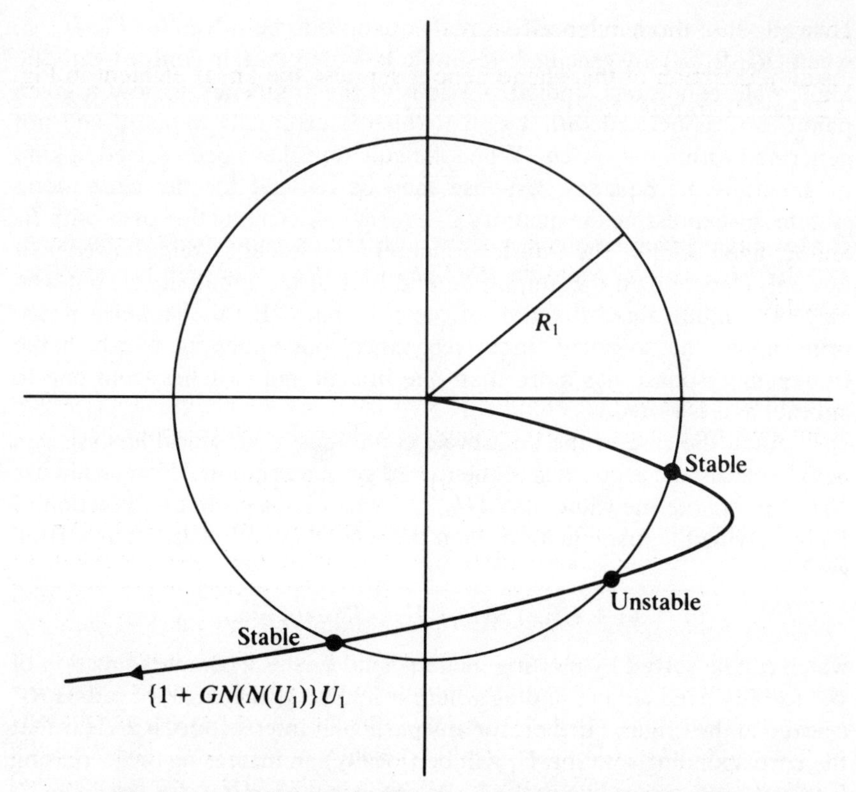

Fig. 3.21 Prediction of stable and unstable forced oscillations.

its output; nevertheless, since the relation between them is frequency-dependent, it can make an effective difference to the form of the frequency response.

Up to this point, we have assumed that $u(t)$ can be approximated by a biased sinusoid at the reference signal frequency, but this is not necessarily true, even if higher harmonics are ignored. It is possible for solutions to contain components at submultiples of the input frequency (subharmonics), or at completely unrelated frequencies, corresponding to the existence of internally generated oscillations, which could have been present even in the absence of a forcing function. In order to apply the harmonic balance method in such cases, it is necessary to go beyond the SIDF and DIDF approximations, by incorporating other terms in the approximants for the signals. Thus, for a subharmonic solution at $1/m$ of the driving frequency ω, we need to retain terms up to the mth harmonic of the fundamental frequency, which is now ω/m. Similarly, if the system is able to support a limit cycle of frequency ω_0, then we must expect to require a combination of terms at frequencies ω and ω_0 (and possibly others) for an adequate representation.

Example 3.8 Jump resonance

As an illustration of this phenomenon, suppose the linear element in Fig. 3.22 has the transfer function

$$G(s) = \frac{K}{s^2+2\zeta s+1}$$

corresponding to a resonant system with unit natural angular frequency and damping ratio ζ, while the nonlinearity is $f(u) = u^3$ and the reference signal is

$$r(t) = R\sin(\omega t)$$

without a bias term, for simplicity. Since the nonlinear function has odd symmetry, we can use the SIDF, namely $N(U) = 3U^2/4$, to obtain

$$\tfrac{3}{4}U^3 + G^{-1}(\mathrm{i}\omega)U = R\exp(-\mathrm{i}\psi)$$

for the parameters of the sinusoidal approximation to $u(t)$. If the expression on the left-hand side is plotted as a function of U, with ω fixed, its locus in the complex plane may intersect the circle centred on the origin, of radius R, either once or three times, depending on the value of ω. Some typical plots are shown in Fig. 3.23, illustrating that, for sufficiently high or low frequency, only one intersection occurs, but over an intermediate range, there are three points of intersection, the outer ones corresponding to stable solutions and the other to an unstable condition. The range of frequency for which this happens will depend, naturally, on the value at which R is set. For a fixed choice of R, the 'frequency response' of the system may perhaps best be represented by plotting, as functions of ω, the magnitude and phase of the quantity $U\exp(\mathrm{i}\psi)/R$, which is analogous to a closed-loop transfer function. Such a plot, of magnitude against frequency, is illustrated in Fig. 3.24, which shows that the function is multivalued over a range (ω_1,ω_2). This has the result that, when the frequency is increased

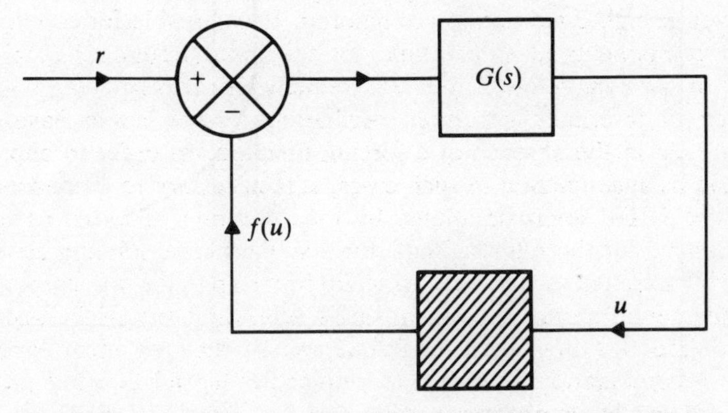

Fig. 3.22 Alternative configuration for a nonlinear feedback loop.

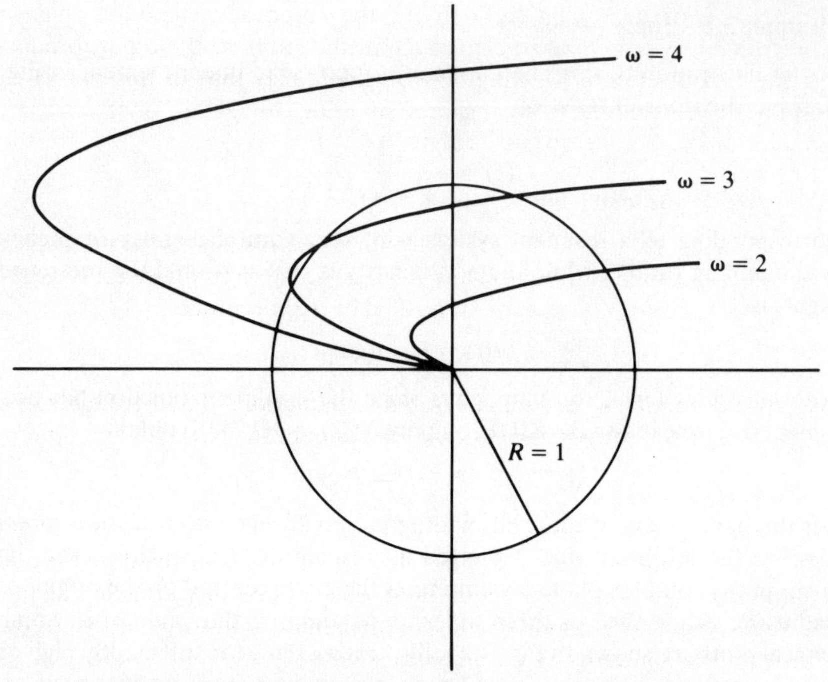

Fig. 3.23 Graphical analysis for Example 3.8 ($\zeta = 0.5$, $K = 5$).

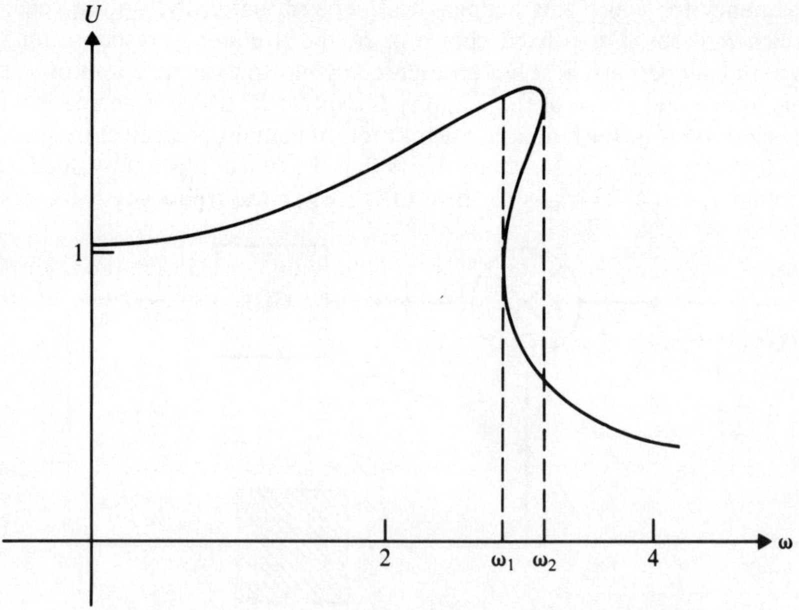

Fig. 3.24 Plot of frequency response for Example 3.8 ($R=1$).

through ω_2, or decreased through ω_1, the response magnitude changes discontinuously from one stable branch of the curve to the other, producing the 'jump' effect. We may note, incidentally, that in this particular case, nothing would have been gained by using the DIDF, since there are no solutions with nonzero bias, provided $K > 0$.

Example 3.9 Subharmonics

The system considered in the previous example turns out also to be capable of sustaining subharmonic oscillations, under some circumstances. If we set $\zeta = 0$, for simplicity, the equation satisfied by $u(t)$ becomes

$$\ddot{u} + u + K u^3 = KR \sin(\omega t)$$

which is sometimes known as Duffing's equation. To investigate the occurence of subharmonic solutions, we assume that $u(t)$ can be approximated by

$$u(t) \simeq U_0 \sin(\tfrac{1}{3}\omega t) + U_1 \sin(\omega t)$$

which is the simplest possibility; the absence of bias is justified by the odd symmetry of the nonlinearity, and the absence of cosine terms by the fact that no phase-shift arises from the dynamics. Ignoring terms of all frequencies other than $\omega/3$ and ω, we then have

$$\begin{aligned} u^3 \simeq {} & (\tfrac{3}{4} U_0^3 - \tfrac{3}{4} U_0^2 U_1 + \tfrac{3}{2} U_0 U_1^2) \sin(\tfrac{1}{3}\omega t) \\ & + (\tfrac{3}{4} U_1^3 + \tfrac{3}{2} U_0^2 U_1 - \tfrac{1}{4} U_0^3) \sin(\omega t) \end{aligned}$$

so that the application of harmonic balancing, with only these two frequencies retained, gives

$$U_0 \left\{ \frac{1}{K}\left(1 - \frac{\omega^2}{9}\right) + \tfrac{3}{4} U_0^2 - \tfrac{3}{4} U_0 U_1 + \tfrac{3}{2} U_1^2 \right\} = 0$$

$$\frac{U_1(1-\omega^2)}{K} + \left(\tfrac{3}{4} U_1^3 + \tfrac{3}{2} U_0^2 U_1 - \tfrac{1}{4} U_0^3 \right) = R$$

whose solution will correspond to a subharmonic oscillation provided that $U_0 \neq 0$. For this to happen, it follows from the first of these equations that we must have

$$\frac{1}{K}\left(1 - \frac{\omega^2}{9}\right) < 0$$

since the quadratic expression involving U_0 and U_1 is positive-definite, and hence $\omega > 3$ for $K > 0$, or $\omega < 3$ for $K < 0$. With the frequency lying in the appropriate range, we then see from the second equation that solutions will exist provided that R is not too large, and that there will then be, in general, two distinct solutions.

In a similar way, we could explore the possibility of oscillations at other submultiples of the driving frequency, the conclusion again being that, under certain conditions, several solutions can exist. Whether or not they are stable, however, is a more delicate question, which appears not to be readily answerable by the harmonic balance method alone.

Example 3.10 Quenching of limit cycles

If an external signal is applied to a nonlinear system capable of self-oscillation, it can interfere with the dynamics so as to alter the parameters of the limit cycle, or even suppress it altogether. To demonstrate this, we take the Van der Pol oscillator of Example 3.6, rearranged in the configuration of Fig. 3.22, so that the forcing signal enters at the input to the linear component. Setting $r(t) = R\sin(\omega t)$, the differential equation for $u(t)$ can be written as

$$\ddot{u} + \epsilon(3u^2-1)\dot{u} + u = \epsilon R\omega\cos(\omega t)$$

which is just Van der Pol's equation with a forcing term. Now, if the driving frequency ω is unrelated to the limit cycle frequency ω_0, it is reasonable to take the approximation

$$u(t) \simeq U_0\sin(\omega_0 t) + U_1\sin(\omega t+\psi)$$

so that

$$u^3 \simeq \tfrac{3}{4}U_0(U_0^2+2U_1^2)\sin(\omega_0 t) + \tfrac{3}{4}U_1(2U_0^2+U_1^2)\sin(\omega t+\psi)$$

following the results of Example 3.3. Hence, applying the harmonic balance method to terms at frequencies ω_0 and ω, we obtain

$$\{G^{-1}(\mathrm{i}\omega_0) + \tfrac{3}{4}(U_0^2+2U_1^2)\}\,U_0 = 0$$

$$\{G^{-1}(\mathrm{i}\omega) + \tfrac{3}{4}(2U_0^2+U_1^2)\}\,U_1\exp(\mathrm{i}\psi) = R$$

where

$$G^{-1}(\mathrm{i}\omega) = \frac{\mathrm{i}}{\epsilon}\left(\omega - \frac{1}{\omega}\right) - 1$$

so that the first equation still gives $\omega_0 = 1$ (assuming $U_0 \neq 0$), but now

$$U_0^2 + 2U_1^2 = \tfrac{4}{3}$$

whence, by substitution into the second equation,

$$\left\{\frac{1}{\epsilon^2}\left(\omega - \frac{1}{\omega}\right)^2 + \left(\frac{9U_0^2}{8} - \frac{1}{2}\right)^2\right\}\left(\frac{2}{3} - \frac{U_0^2}{2}\right) = R^2$$

after eliminating ψ. It is thus clear that, for any given value of ω, there is a bound on the magnitude of R for which a solution of this type is possible; if this bound is exceeded, then the only possibility is that $U_0 = 0$, in which case the component of the solution at the limit cycle frequency disappears entirely. This is sometimes expressed by saying that, for a certain range of values of the input amplitude and frequency, the limit cycle is 'quenched'. In that case, the only remaining component, in this approximation, is at the frequency of the driving signal, and its parameters can be calculated by applying harmonic balance at this frequency alone.

3.5 Exercises

1 Calculate the SIDF for the backlash function shown in Fig. 3.25, assuming the input amplitude $U \geqslant a$.

2 For the feedback system of Fig. 3.12, with constant reference input, the nonlinearity being a dead zone as in Fig. 3.6, and the linear element having the transfer function

$$G(s) = \frac{K}{s(s+1)^2}$$

show that an unstable limit cycle is predicted, using the SIDF approximation, if K is large enough.

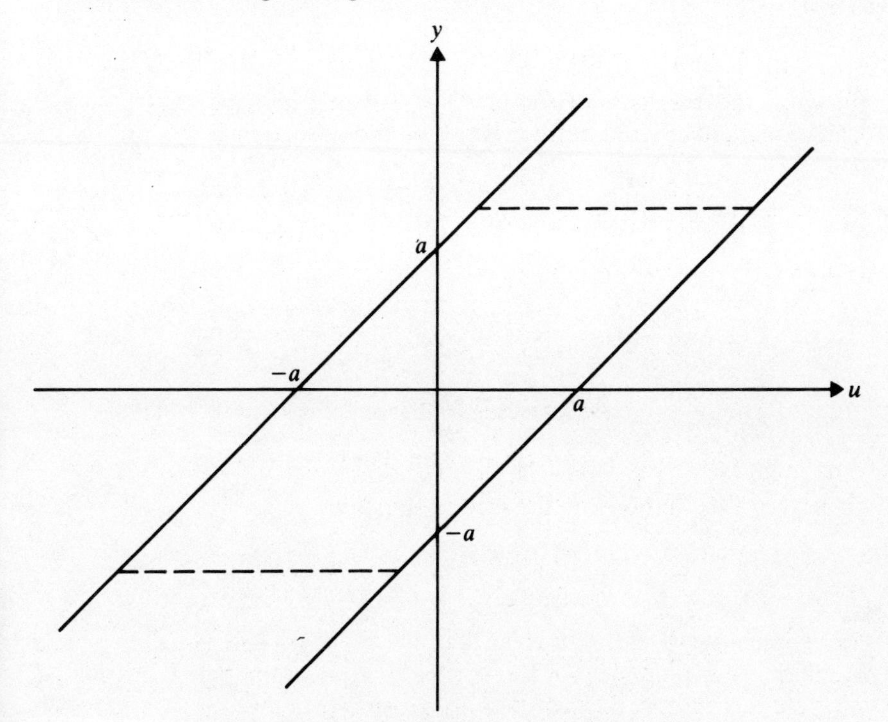

Fig. 3.25 A nonlinearity representing backlash.

3 Apply the DIDF approximation to the system of Fig. 3.12, when the nonlinearity is the quadratic function $f(u) = u^2$, the transfer function of the linear element is

$$G(s) = \frac{1-\lambda s}{s^2+2\zeta s+1}$$

and the reference input r is constant. Show how the limit cycle predictions depend on the value of r and the positive constants λ and ζ.

4 Investigate the possibility of jump resonance in Fig. 3.12, with the cubic nonlinearity $f(u) = u^3$ and the transfer function

$$G(s) = \frac{1+s}{s^2}.$$

4 PIECEWISE - LINEAR MODELS

Many nonlinearities encountered in practice can usefully be approximated by piecewise-linear relations, and we have already noted such examples as the saturation, dead zone, relay and backlash characteristics. The most important feature of this type of model is that it enables the state space to be divided into a number of distinct regions, in each of which the system's dynamical behaviour can be analysed by linear techniques, with the solutions for different regions then being matched together at the boundaries. Clearly, any particular trajectory must either remain permanently in one region, in which case it is equivalent to a trajectory of the corresponding linear system, or cross repeatedly from some region to another; in the latter case, it can be characterised by a discrete sequence of values, which are the coordinates of the points where the boundaries are crossed. Consequently, dynamical properties such as periodicity and stability can be investigated by studying the behaviour of these sequences, which are generated by the mapping from one crossing point to the next, induced by the linear dynamics appropriate for the intervening region. This technique, known as the 'point transformation' method, is particularly useful in deriving conditions for the existence of limit cycles, especially in second-order systems, where only one coordinate is needed to label the points on a boundary. It should be noted, also, that this approach does not depend on the piecewise-linear function being single-valued, and can just as well be applied, for instance, to a system containing a relay device with hysteresis, where the boundaries would correspond to the switching points. Systems of this kind have, in effect, a slightly higher 'order' than dynamical considerations would suggest; thus, even if there are only two (continuous) state variables, an extra (discrete) parameter is required to label the branches of the multivalued relation, so that the phase space is no longer strictly two-dimensional.

4.1 The point transformation method

To illustrate the procedure, let us first consider the simple case of a second-order system, described by two different sets of linear equations in different regions, separated by a boundary as shown in Fig. 4.1. Labelling the points on the boundary by a parameter p, suppose a trajectory starts at $p = p_1$ and goes into region I. By solving the equations for this region, we can find where (if at all) the trajectory again reaches the boundary; calling the coordinate of this point p_2, we then have a functional relationship between p_1 and p_2. Next, suppose that the trajectory goes from p_2 into region II, and returns to the boundary again at $p = p_3$, so that, after solving the appropriate dynamical equations, we can obtain p_3 as a function of p_2. Now, it is clear from Fig. 4.1 that the condition for a limit cycle to exist is $p_3 = p_1$, while, by eliminating p_2, it follows that there is a functional dependence of the form $p_3 = \phi(p_1)$. We can thus solve the limit cycle condition graphically by plotting this function as in Fig. 4.2 (known as a Lemeré diagram), and finding where it intersects the line $p_3 = p_1$, if it does so. This procedure can still be used even if the elimination of p_2 is impracticable, provided that p_1 and p_3 can both be expressed as functions of p_2 (or some other parameter), thus enabling the graph to be plotted. Moreover, the stability of the limit cycle can also be determined from the same diagram since, if Δp_1 and Δp_3 denote small deviations from a point of intersection, we have

$$\Delta p_3 = \phi'(p_1)\, \Delta p_1$$

where ϕ' is the derivative of ϕ. Thus, the magnitude of the deviation will grow or decrease with successive iterations, according as $|\phi'(p_1)|>1$ or <1;

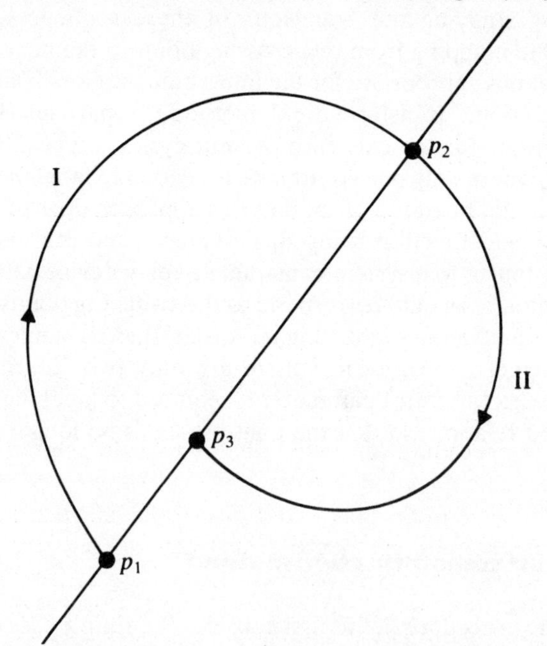

Fig. 4.1 Phase plane with two regions separated by a boundary.

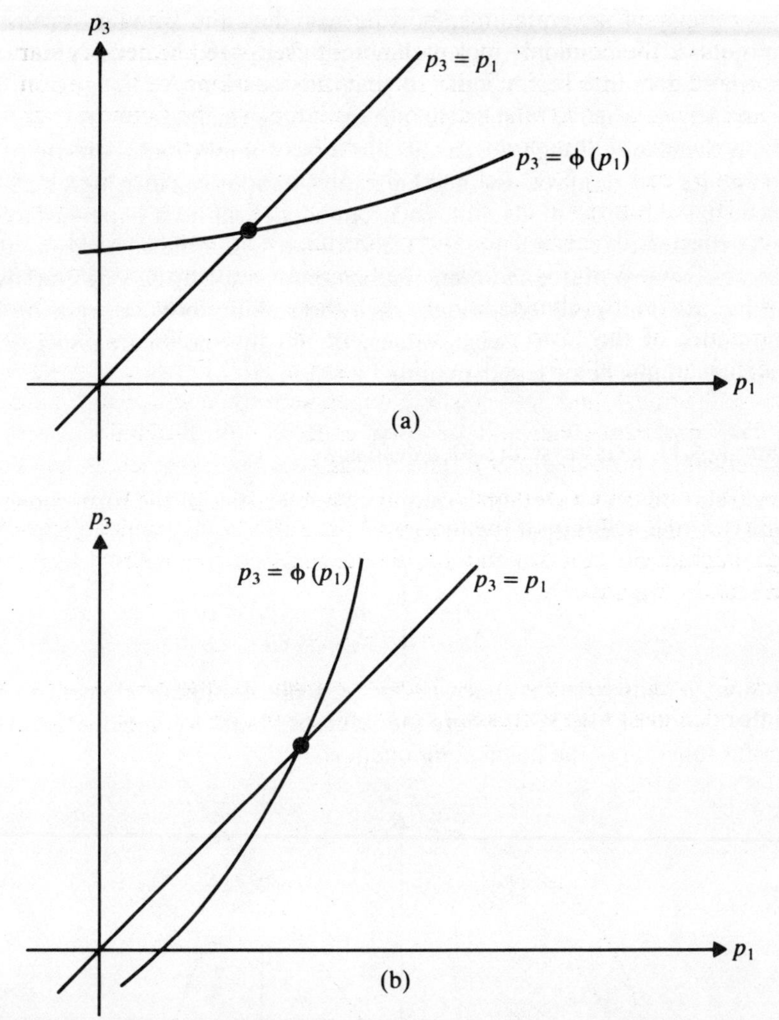

Fig. 4.2 Lemeré diagram for: (a) stable limit cycle; (b) unstable limit cycle.

that is to say, the limit cycle will be stable if the slope of the graph at the intersection point is less than unity in magnitude, unstable if it is greater.

The above procedure can readily be extended to the case of a two-dimensional system with multiple boundaries, and also to the slightly more general situation with multivalued relations, provided that the dynamical part of the system is still of second order. One simply labels successive boundary crossing points by parameters $(p_1, p_2, \ldots, p_m)$, so that the limit cycle condition becomes $p_m = p_1$, where p_m is given as a function of p_1 by solving the linear differential equations for the various regions traversed, and then proceeds as before. In some cases, it may be possible to exploit the symmetry of the equations, in order to reduce the amount of calculation required; for example, only half a limit cycle need be considered, in a case where the other half is bound to be its mirror image.

With systems of higher dimension, the principles of the method remain the same, though the details are more complicated, because more parameters are required for labelling the boundaries; in general, if the system has nth-order dynamics, each boundary will be a hypersurface of $(n-1)$ dimensions, needing $(n-1)$ coordinates. Graphical analysis thus becomes inappropriate, and numerical methods must usually be employed to solve the limit cycle conditions; also, the stability now depends on the magnitudes of the $(n-1)$ eigenvalues, of the matrix relating successive small deviations, being less than unity.

Example 4.1 Relay system with a dead zone

Some relay devices have input–output characteristics of the form shown in Fig. 4.3, where the output is given by

$$f(u) = \begin{cases} h & u > b \\ 0 & |u| < b \\ -h & u < -b \end{cases}$$

for input u. Suppose that such a device is the nonlinear element in the feedback loop of Fig. 3.12, where the reference input r is constant, and the transfer function of the linear component is

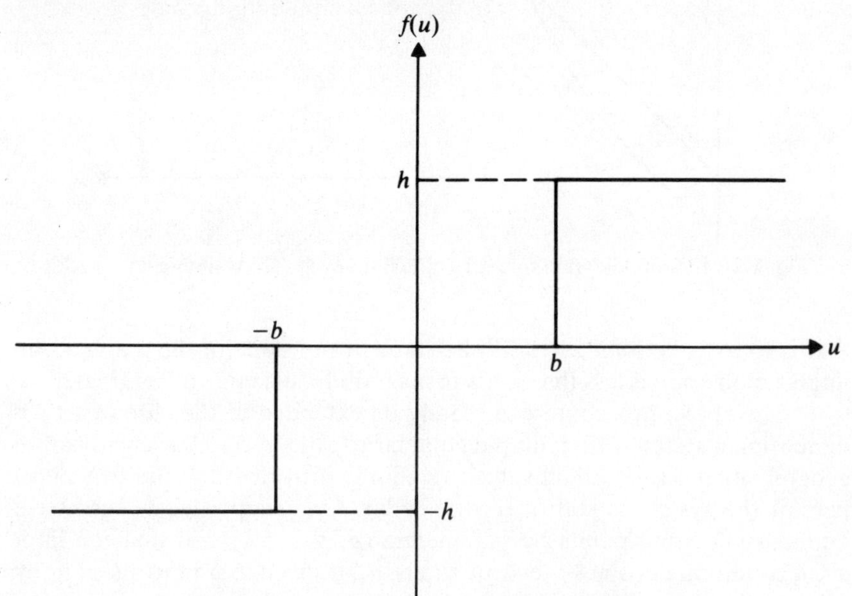

Fig. 4.3 Relay characteristic with a dead zone.

$$G(s) = \frac{1 - Ks}{s(s+1)}$$

for some positive constant K. Expressing this in partial fractions as

$$G(s) = \frac{1}{s} - \frac{(1+K)}{s+1}$$

we see that a state-space representation can be set up in the form

$$\begin{aligned} \dot{x}_1 &= f(u) \\ \dot{x}_2 &= -x_2 + (1+K)\,f(u) \\ y &= c + x_1 - x_2 \end{aligned}$$

where c is an arbitrary constant which can be added to x_1 without altering the differential equations. Choosing $c = r$, we have

$$u = r - y = x_2 - x_1$$

so that the constant reference input disappears from the equations; the fact that this can be done is a consequence of the integrator term in $G(s)$. The two-dimensional state space is illustrated in Fig. 4.4, where the switching boundaries ($x_1 - x_2 = \pm b$) are parametrised by

$$(x_1,x_2) = \begin{cases} (p,p-b) & u = -b \\ (-q,b-q) & u = b \end{cases}$$

Now, consider a trajectory starting at $p = p_1$ when $t = 0$ and going into the region $u = -b$, so that the dynamical equations become

$$\begin{aligned} \dot{x}_1 &= -h \\ \dot{x}_2 + x_2 &= -(1+K)h \end{aligned}$$

with the solutions

$$\begin{aligned} x_1 &= p_1 - ht \\ x_2 &= \{(1+K)h + p_1 - b\}\exp(-t) - (1+K)h \end{aligned}$$

reaching the boundary again at $p = p_2$ when $t = \tau$. We thus have

$$\begin{aligned} h\tau &= p_1 - p_2 \\ \{(1+K)h + p_1 - b\}\exp(-\tau) &= (1+K)h + p_2 - b \end{aligned}$$

so that, on eliminating τ,

$$\exp\left(\frac{p_1 - p_2}{h}\right) = \frac{(1+K)h + p_1 - b}{(1+K)h + p_2 - b}.$$

Next, suppose the trajectory then passes into the region $|u| < b$, where the state-space equations give

$$\dot{x}_1 = 0, \qquad \dot{x}_2 = -x_2$$

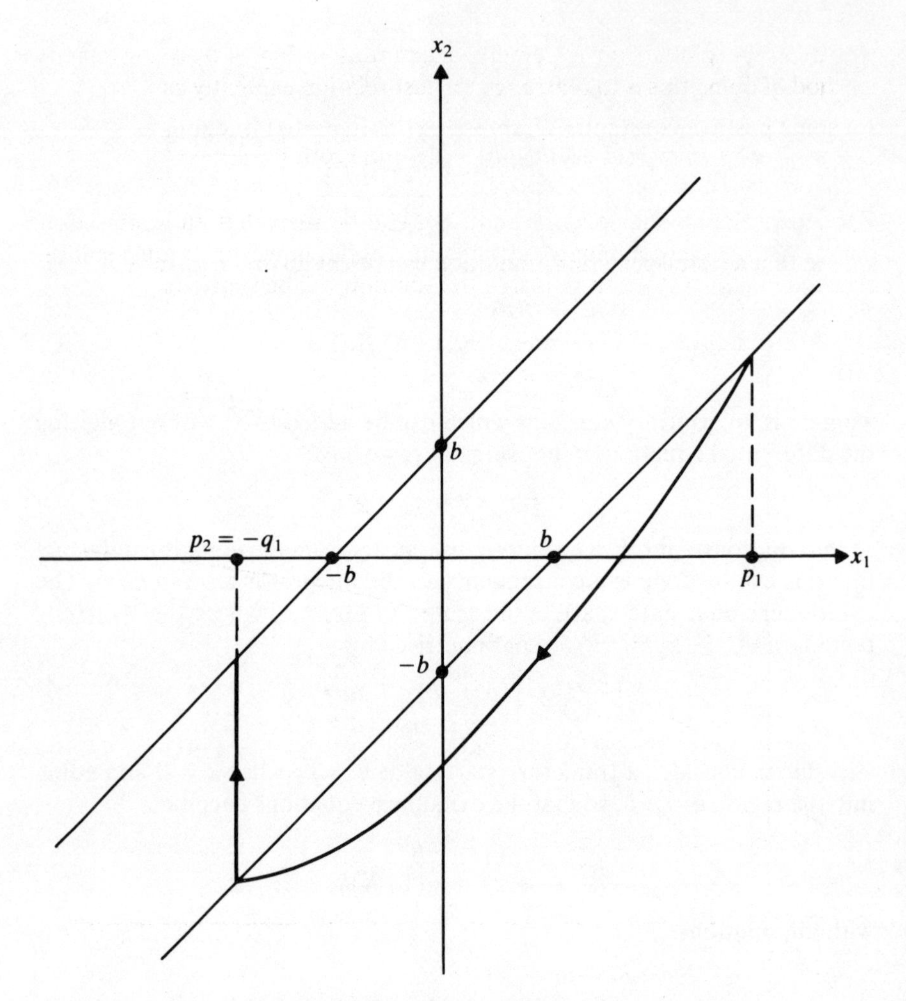

Fig. 4.4 State space for Example 4.1.

so that x_1 remains constant at p_2. This section of the trajectory thus lies along the line $x_1 = p_2$ and will eventually reach the other boundary ($u = b$) at $q = q_1$, where $q_1 = -p_2$, provided that $p_2 < -b$; otherwise, it would terminate on reaching $x_2 = 0$. From the fact that the system equations are unaltered by reversing the signs of x_1, x_2 and u, the limit cycle condition can be taken as

$$q_1 = p_1$$

where, by eliminating p_2, we have

$$\exp\left(\frac{p_1+q_1}{h}\right) = \frac{(1+K)h - b + p_1}{(1+K)h - b - q_1}$$

which can be plotted on a Lemeré diagram as in Fig. 4.5; a convenient method of doing this is to rearrange the last relation explicitly as

$$q_1 - p_1 = 2\{(1+K)h - b\} - (p_1+q_1)\coth\left(\frac{p_1+q_1}{2h}\right).$$

For appropriate values of h, b and K, it can be seen that an intersection point exists, corresponding to a stable limit cycle; however, it must still be checked that $q_1 > b$ at this point, or the solution will be spurious.

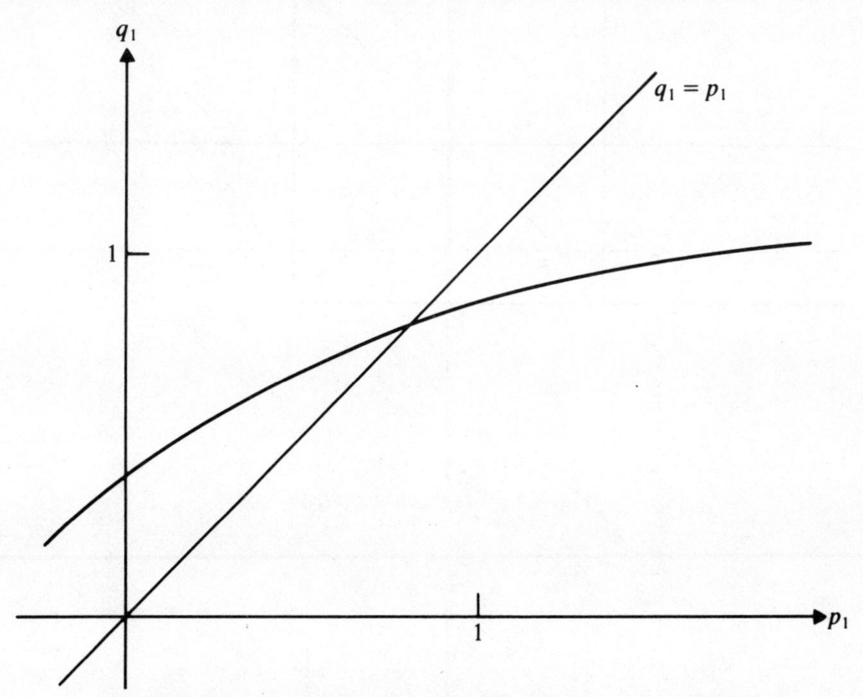

Fig. 4.5 Lemeré diagram for Example 4.1 (h=1, b=0.6, K=0.8).

Example 4.2 A chaotic system

A simple system, which can exhibit chaotic behaviour, is described by the differential equation

$$\ddot{u} - 2\alpha\dot{u} + u = f(u)$$

where α is a constant between 0 and 1, and $f(u)$ is the two-valued relay characteristic shown in Fig. 4.6, corresponding to a device sometimes known as a 'toggle'. Introducing state variables $x_1 = u$, $x_2 = \dot{u}$, as in Fig. 4.7, let us consider a trajectory starting when the toggle has just switched to the state with $f(u) = 1$, that is to say, at

$$(x_1,x_2) = (1,p)$$

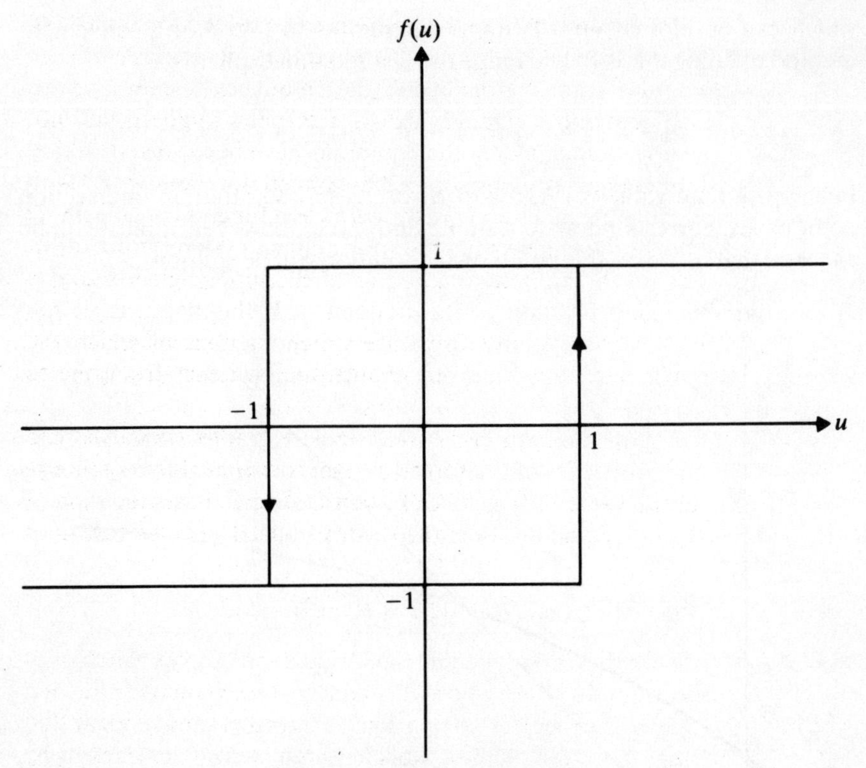

Fig. 4.6 Toggle characteristic.

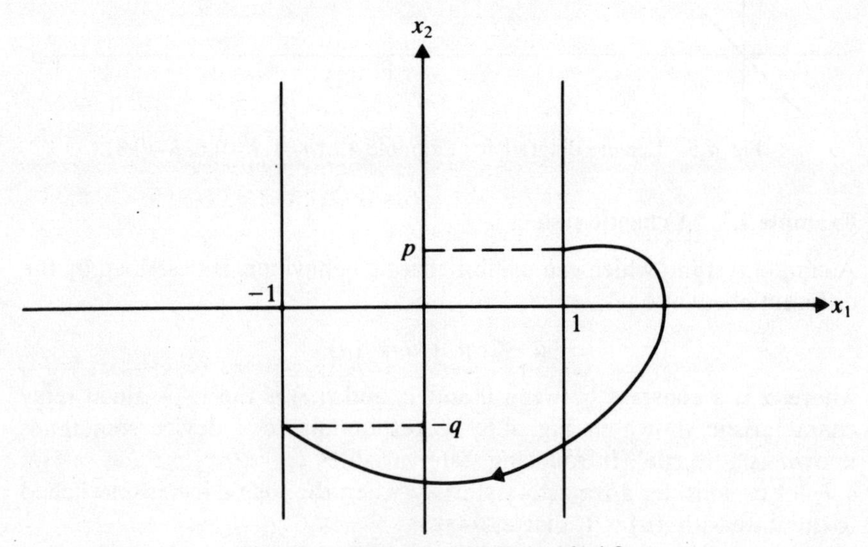

Fig. 4.7 Phase plane for Example 4.2.

where $p > 0$, as u must be increasing through 1. Solving the differential equations, we then obtain

$$x_1 = u = 1 + \frac{p}{\beta}\exp(\alpha t)\sin(\beta t)$$

$$x_2 = \dot{u} = \frac{p}{\beta}\exp(\alpha t)\{\alpha\sin(\beta t) + \beta\cos(\beta t)\}$$

where $\beta = \sqrt{(1 - \alpha^2)}$. The solution will continue to evolve until u first reaches the value -1, when the toggle switches to the state with $f(u) = -1$. Denoting the point at which this happens by

$$(x_1, x_2) = (-1, -q)$$

we thus have

$$\exp(\alpha\tau)\sin(\beta\tau) = -2\beta/p$$

$$\exp(\alpha\tau)\cos(\beta\tau) = (2\alpha - q)/p$$

where τ is the time elapsed from the start. If the relation between p and q obtained in this way is denoted by

$$q = \phi(p)$$

it follows, from the symmetry of the system, that all subsequent switching points are generated by repeated application of the function $\phi(\cdot)$. In order to examine the form of this function, we obtain, from the above equations,

$$\exp(\alpha\tau) = \frac{\sqrt{(q^2 - 4\alpha q + 4)}}{p}$$

whence

$$\cos(\beta\tau) = \frac{2\alpha - q}{\sqrt{(q^2 - 4\alpha q + 4)}}$$

where we also know, since p is positive, that $\sin(\beta\tau) < 0$, and hence

$$\beta\tau = 2k\pi - \arccos\left(\frac{2\alpha - q}{\sqrt{(q^2 - 4\alpha q + 4)}}\right)$$

for some positive integer k, taking the usual convention that the inverse cosine lies between 0 and π. We can thus plot p against q by using this last result together with

$$p = \sqrt{(q^2 - 4\alpha q + 4)}\exp(-\alpha\tau)$$

giving an infinitely-many-valued relation with one branch corresponding to each choice of k. When inverting this relation to give q as a function of p, we have always to take the smallest possible k, since this gives the lowest value of τ, corresponding to the first time that the trajectory hits the switching boundary. This results in a discontinuous function, as illustrated by the Lemeré diagram of Fig. 4.8, for a particular value of α.

It can be seen that the gradient of the function $\phi(p)$ is always greater than unity, so that any limit cycle, corresponding to an intersection between this function, or one of its iterates, and the line $q = p$, must be unstable. Moreover, for sufficiently small α, there exists a number $\mu > 0$, such that $p \leqslant \mu$ implies $q \leqslant \mu$, and so any trajectory which starts with p small enough will remain bounded. Consequently, since the only equilibrium points ($u = \pm 1$) are also unstable, it follows that there is a region within which trajectories remain permanently, without converging on to any stable equilibrium or periodic solution. The resulting motion is thus bounded but chaotic, with almost all trajectories being non-periodic, and the periodic ones unstable.

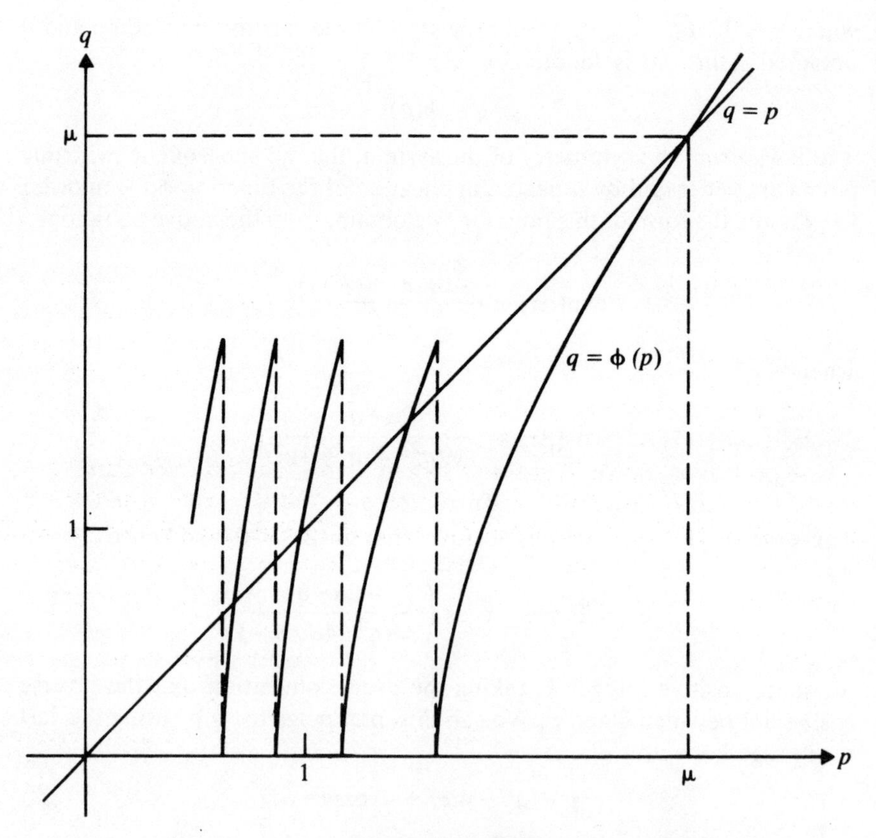

Fig. 4.8 Lemeré diagram for Example 4.2 ($\alpha = 0.05$).

Example 4.3 The stick–slip phenomenon

In some systems whose equations contain discontinuous functions, it is necessary to examine in detail what happens on the switching boundaries, in order to get a complete picture of the dynamical behaviour. A simple mechanical example consists of a moveable mass, restrained by a spring but in frictional contact with a steadily moving belt, as illustrated in Fig. 4.9. Taking unit values for the mass and spring constant, the position coordinate y obeys the equation

$$\ddot{y} + y = f$$

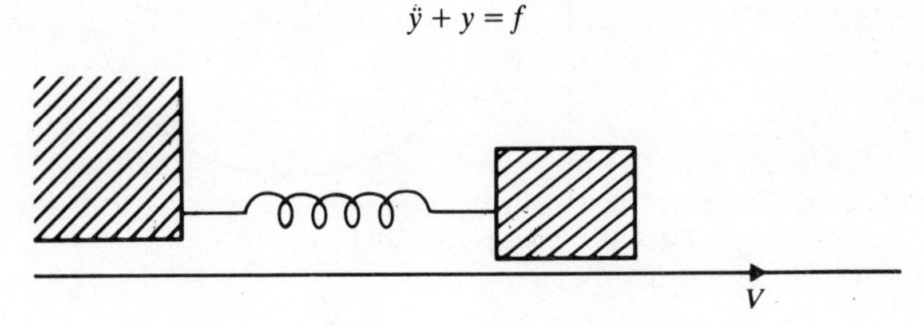

Fig. 4.9 Tethered mass resting on a moving belt.

where the frictional force f depends on whether the mass is sliding over the belt (slip) or adhering to it (stick). For the slipping mode, we assume Coulomb friction applies, with

$$f = \text{sgn}(V - \dot{y})$$

where V is the belt velocity, while, in the sticking mode, when evidently $\dot{y} = V$, the friction can take any value between $\pm F$ (the maximum static friction). Using $x_1 = y$ and $x_2 = \dot{y}$ as state variables, we see that the trajectories can take three possible forms: for $x_2 < V$, they consist of circles (or circular arcs) centred on the point (1,0); for $x_2 > V$, arcs of circles centred on $(-1,0)$; and also segments of the line $x_2 = V$. The resulting phase portrait is shown in Fig. 4.10, where it has been assumed that $F > 1$, that is to say, the static frictional force is capable of exceeding the (dynamic) Coulomb friction. It is clear that, in the range $-1 < x_1 < 1$, trajectories approach the boundary $x_2 = V$ from both sides, so that the subsequent motion must be assumed to remain on this line, with x_1 increasing, until it exceeds unity. On physical grounds, it is to be expected that this will in fact continue until $x_1 = F$, where the friction reaches its maximum static value, and that the trajectory will then pass into the region $x_2 < V$, returning to the boundary at $(2-F,V)$. We thus deduce that the system possesses a limit cycle, lying partly in $x_2 < V$ (slip mode) and partly along $x_2 = V$ (stick mode). Besides this, we also see that, in the slipping regime, undamped oscillations, of any amplitude less than V, can occur, corresponding to circles which do not intersect the boundary.

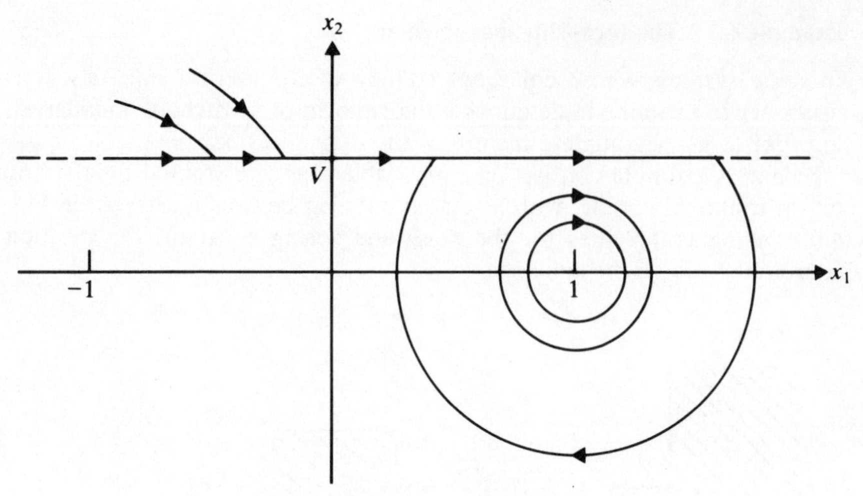

Fig. 4.10 Phase portrait for Example 4.3.

4.2 Variable-structure systems

A special kind of piecewise-linear system arises from the implementation of a type of control scheme known as variable-structure control. This is a modification of the conventional feedback design, where the control input is synthesised as a linear combination of the state variables, with fixed coefficients; in a variable-structure system, the coefficients are switched from one set of values to another, when the trajectories cross certain hyperplanes in the state space, or equivalently, when some combinations of state variables vanish. There are several possible advantages in doing this: in the first place, the extra structure can potentially be exploited to improve performance features, such as speed of response, and also, in some cases, to make the system insensitive to parameter variations; moreover, when attempting to regulate several output variables simultaneously, it may well be beneficial to adopt a 'competitive' strategy of changing the control law in accordance with their relative magnitudes, thus leading naturally to a variable-structure design.

The simplest case to consider is a plant which acts as a double integrator, so that its input u and output y are related by

$$\ddot{y} = u.$$

Suppose we wish to make the output follow a reference signal r, which is either constant or uniformly varying, so that $\dot{r}$ is constant. If we introduce state variables

$$x_1 = y - r, \qquad x_2 = \dot{y} - \dot{r},$$

the state-space equations become

$$\dot{x}_1 = x_2$$
$$\dot{x}_2 = u$$

sine $\ddot{r}$ vanishes, and our objective can be achieved by using the feedback law

$$u = -f_1x_1 - f_2x_2$$

for any positive constants (f_1, f_2), as the closed-loop system is then asymptotically stable. For the reasons mentioned above, however, we may prefer to adopt a variable-structure scheme, in which the values of the feedback coefficients are changed when certain lines in the phase plane are crossed. A typical choice would be to take the 'switching lines' as

$$x_1 = 0, \qquad \lambda x_1 + x_2 = 0$$

for some constant λ, as illustrated in Fig. 4.11, and to switch the coefficients according to the rules

$$f_j = \begin{cases} g_j & \sigma x_1 > 0 \\ h_j & \sigma x_1 < 0 \end{cases}$$

for $j = 1,2$, where g_j and h_j are constants, and

$$\sigma = \lambda x_1 + x_2.$$

It can be seen that this system, though not linear, still possesses a sort of quasi-linearity, in that the dynamical equations are unaltered if the state variables are re-scaled by a common factor. As a result of this 'scale

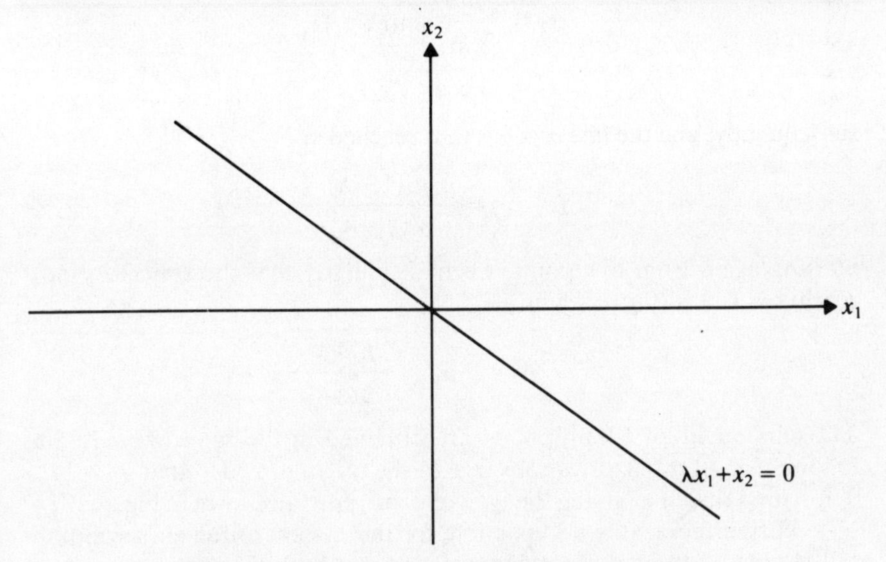

Fig. 4.11 Phase plane for variable-structure system.

invariance', limit cycling is unlikely and the system may be simply classified as either stable or unstable, just as in the linear case. To determine the stability properties, the most direct approach is to use the point transformation method, which will work provided that the trajectories keep on crossing the switching lines. On the other hand, for some choices of feedback gains, trajectories may approach the line $\sigma = 0$ from both sides, and therefore subsequently remain on this line, so that

$$\dot{x}_1 = -\lambda x_1.$$

In this mode, which is known as 'sliding' motion, the behaviour is evidently asymptotically stable if $\lambda > 0$, even though the system as a whole need not be stable.

Example 4.4 Asymptotic stability without damping

In the system corresponding to Fig. 4.11, let us consider the special case $f_2 = 0$, so that, for a fixed $f_1 > 0$, the feedback loop has the dynamics of an undamped harmonic oscillator. With a variable-structure controller, however, we have

$$f_1 = \begin{cases} g_1 & \sigma x_1 > 0 \\ h_1 & \sigma x_1 < 0 \end{cases}$$

which allows a variety of dynamical behaviour. To apply the point transformation method, suppose that a trajectory starts at the point $(0, p_1)$ when $t = 0$, so that

$$x_1 = \frac{p_1}{\sqrt{g_1}} \sin(t\sqrt{g_1})$$

$$x_2 = p_1 \cos(t\sqrt{g_1})$$

subsequently, and the line $\sigma = 0$ is thus reached at

$$(x_1, x_2) = \frac{p_1(1, -\lambda)}{\sqrt{(g_1+\lambda^2)}}.$$

Starting again from this point, we find similarly that the trajectory next reaches $x_1 = 0$ at $(0, p_2)$, where

$$p_2 = -p_1 \sqrt{\frac{h_1+\lambda^2}{g_1+\lambda^2}}$$

and subsequent switching points are obtained in the same way. It thus follows that all trajectories converge to the origin provided that $g_1 > h_1 > 0$, for any value of λ, giving the type of phase portrait shown in Fig. 4.12.

Furthermore, it is also possible for this system to remain asymptotically stable when h_1 is negative, provided that both g_1 and λ are positive.

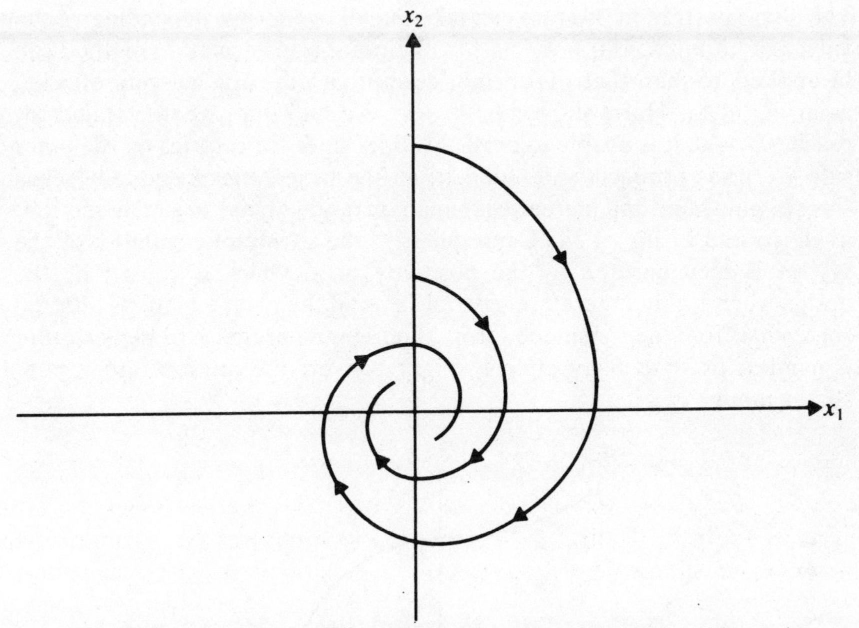

Fig. 4.12 Trajectories of Example 4.4 (g_1=2, h_1=0.5, λ=0).

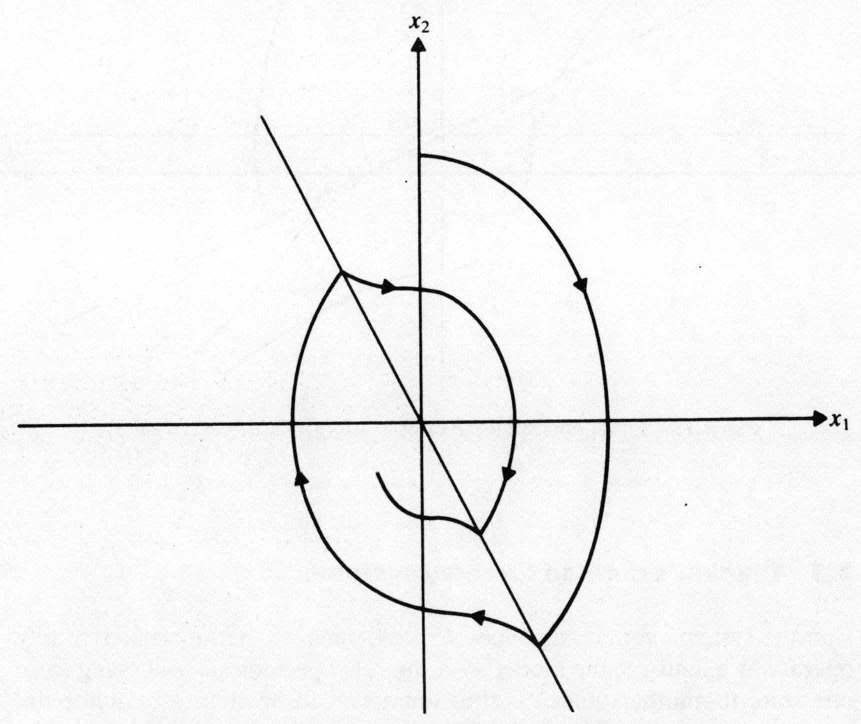

Fig. 4.13 Trajectory of Example 4.4 (g_1=2, h_1=−1, λ=2).

The phase portrait in this case can take one of two forms, depending on the values of h_1 and λ. If $0 > h_1 > -\lambda^2$, the point transformation method can be applied, to show that all trajectories approach the origin asymptotically, as in Fig. 4.13. However, for $h_1 < -\lambda^2$, we find that, when a trajectory reaches $\sigma = 0$, it is unable to cross this line, since trajectories on the other side are also approaching it. Indeed, all the trajectories eventually reach this switching line and thereafter remain on it, giving rise to sliding motion, as illustrated in Fig. 4.14. Consequently, the asymptotic stability of the system is then ensured by the positivity of λ, which is chosen by the designer, and would therefore still hold even if the plant equations differed somewhat from their assumed form, although the presence of higher-order dynamics, or time-delay effects, might convert the motion into a non-sliding mode.

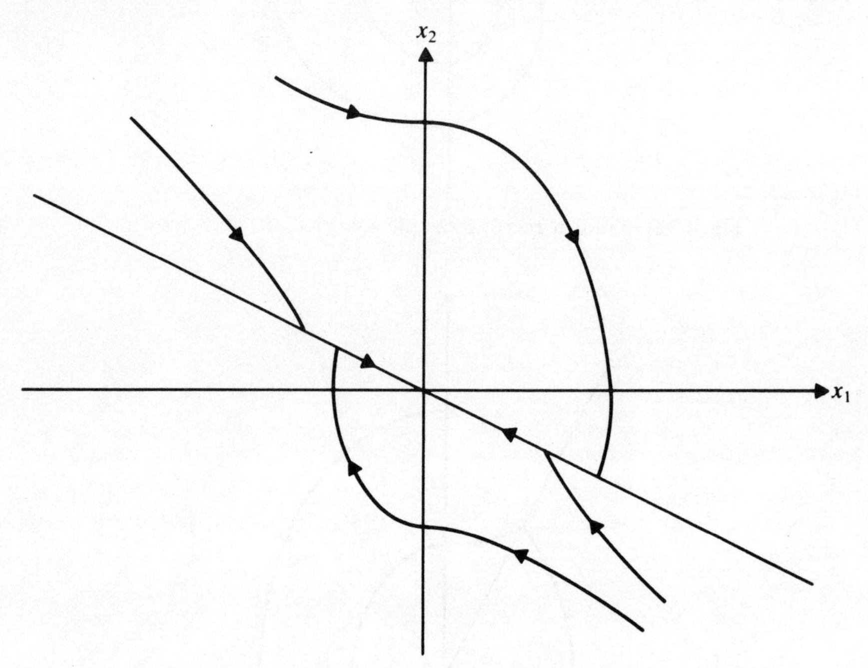

Fig. 4.14 Phase portrait for Example 4.4 ($g_1=2$, $h_1=-1$, $\lambda=0.5$).

4.3 Tsypkin's method for relay systems

Control systems containing relay devices, such as thermostats, typically operate in a limit cycling mode, with the relay periodically switching from one state to another, and it is thus important to be able to calculate the parameters of these oscillations. Since the output of the relay is necessarily a piecewise-constant signal, this fact may as well be exploited, and one method of doing so, using a frequency-domain formulation, is due to Tsypkin; the equivalent time-domain version, based on the point transformation method in the state space, was developed by Hamel. In Tsypkin's approach, one begins by assuming that the relay output has a simple periodic form, which is then expressed as a Fourier series; because the rest of the system is linear, the Fourier expansion of the relay input signal can then be found, and the appropriate switching conditions imposed, leading to a set of equations from which the limit cycle parameters can be computed.

The simplest piecewise-constant periodic signal to analyse is the square pulse train shown in Fig. 4.15, where $v(t)$ switches from 0 to 1 at $t = 0$, back to 0 at $t = \tau$, back again to 1 at $t = T$, and so on. Using the standard expressions for Fourier coefficients given in Chapter 3, we obtain

$$v(t) = \frac{\tau}{T} + \frac{1}{\pi} \sum_{k=1}^{\infty} \frac{\sin(k\omega t) - \sin\{k\omega(t-\tau)\}}{k}$$

where $\omega = 2\pi/T$, the angular frequency corresponding to period T. It will be convenient to rewrite this in complex notation as

$$v(t) = \frac{1}{2\pi \mathrm{i}} \sum_{k=-\infty}^{\infty} \frac{1-\exp(-\mathrm{i}k\omega\tau)}{k} \exp(\mathrm{i}k\omega t)$$

since, if this signal is fed into a linear dynamical system, the output can be evaluated by multiplying each term by the transfer function at the appropriate frequency; we note, incidentally, that the term with $k = 0$ is correctly given by taking the limit of the coefficient expression as '$k \to 0$'. For the investigation of limit cycles in relay systems, it may be necessary to deal with considerably more complicated signals than this, but they can be similarly analysed and, indeed built up from combinations of the same basic form.

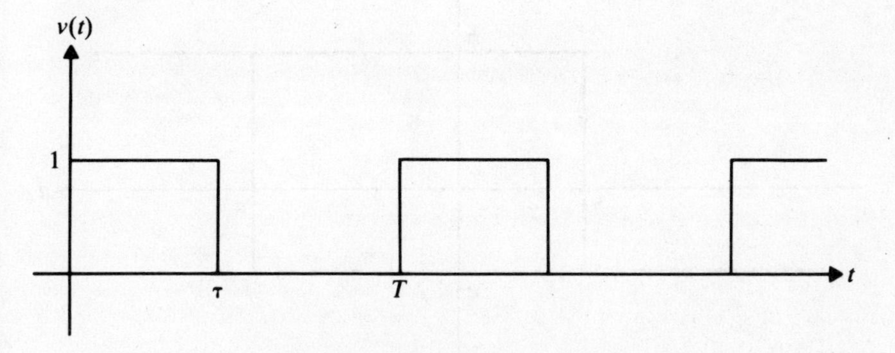

Fig. 4.15 A train of square pulses.

In order to derive the Tsypkin conditions for a limit cycle, we define a complex function $\Lambda(\theta,\omega)$ by setting

$$\mathrm{Re}\Lambda(\theta,\omega) = \frac{2}{\pi} \sum_{k=1}^{\infty} \{\mathrm{Re}G(\mathrm{i}k\omega)\cos(k\theta) + \mathrm{Im}G(ik\omega)\sin(k\theta)\}$$

$$\mathrm{Im}\Lambda(\theta,\omega) = \frac{2}{\pi} \sum_{k=1}^{\infty} \frac{\mathrm{Im}G(ik\omega)\cos(k\theta) - \mathrm{Re}G(ik\omega)\sin(k\theta)}{k}$$

where θ and ω are real parameters, and $G(s)$ is the transfer function of the linear element in the feedback loop. To avoid problems with the convergence and continuity of these expressions, we assume, for the time being, that $G(s)$ tends to zero faster than $1/s$ as $s \to \infty$, though this assumption will be relaxed later. Now, suppose the system is configured as in Fig. 3.12, with the nonlinear element being a two-state relay, whose output $f(u)$ takes the values h and d in the upper and lower states, respectively, so that $h>d$, and switches up and down when $u = a$ and $u = b$, respectively, where $a \geqslant b$, as shown in Fig. 4.16. With the reference input r held constant, the steady mode of operation of the system is assumed to be an oscillation with period T, where the relay spends alternate intervals of duration τ and $(T-\tau)$, respectively, in its upper and lower states, so that its output is given by

$$f(u) = d + (h-d)\, v(t)$$

in terms of the function $v(t)$ defined above. This signal becomes the input to the linear element, whose output $y(t)$ is related to the relay input by $u = r-y$, so that the switching conditions are

$$y(0) = r - a, \qquad \dot{y}(0) \leqslant 0,$$
$$y(\tau) = r - b, \qquad \dot{y}(\tau) \geqslant 0,$$

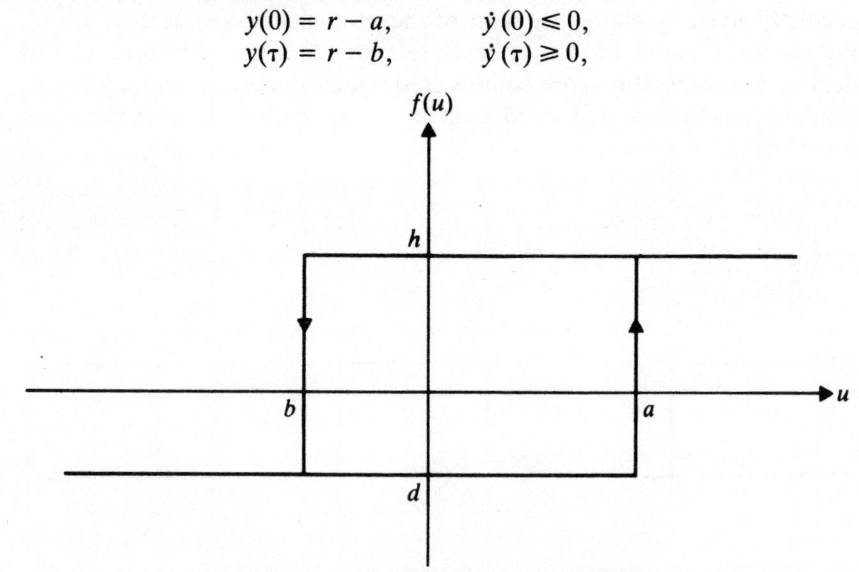

Fig. 4.16 Unsymmetrical relay characteristic.

in order for $u(t)$ to be passing through the switching points in the correct directions at the appropriate instants. Using the Fourier expansion for $v(t)$ and the definition of $\Lambda(\theta,\omega)$, we find, after some algebra,

$$y(0) = G(0)\left(d+(h-d)\frac{\tau}{T}\right)+\frac{h-d}{2}\,\mathrm{Im}\{\Lambda(0,\omega)-\Lambda(\omega\tau,\omega)\}$$

$$y(\tau) = G(0)\left(d+(h-d)\frac{\tau}{T}\right)+\frac{h-d}{2}\,\mathrm{Im}\{\Lambda(-\omega\tau,\omega)-\Lambda(0,\omega)\}$$

with $\omega = 2\pi/T$, and also, similarly,

$$\dot{y}(0) = \frac{(h-d)\omega}{2}\,\mathrm{Re}\{\Lambda(0,\omega)-\Lambda(\omega\tau,\omega)\}$$

$$\dot{y}(\tau) = \frac{(h-d)\omega}{2}\,\mathrm{Re}\{\Lambda(-\omega\tau,\omega)-\Lambda(0,\omega)\}$$

though these latter expressions, for $\dot{y}$, are only valid under our assumption that $sG(s) \to 0$ as $s \to \infty$. Otherwise, $\dot{y}(t)$ may be discontinuous at the switching instants, and the conditions actually required are then

$$\dot{y}(0-) \leqslant 0, \qquad \dot{y}(\tau-) \geqslant 0,$$

whereas the values calculated from the Fourier series are the averages across the discontinuities, namely

$$\dot{y}(0) = \frac{\dot{y}(0-)+\dot{y}(0+)}{2}$$

$$\dot{y}(\tau) = \frac{\dot{y}(\tau-)+\dot{y}(\tau+)}{2},$$

using an obvious notation for the limits. Hence, if

$$sG(s) \to \rho$$

as $s \to \infty$, for some constant ρ, so that the signal $\{\dot{y}-\rho f(u)\}$ is continuous, being related to $f(u)$ by the transfer function $\{sG(s)-\rho\}$, we can compute the required quantities by using the equations

$$\dot{y}(0+)-\dot{y}(0-) = \dot{y}(\tau-)-\dot{y}(\tau+) = (h-d)\rho$$

which follow from the known discontinuities in $f(u)$. As a result, the inequality conditions can be combined into

$$\mathrm{Re}\{\Lambda(0,\omega)-\Lambda(\pm\omega\tau,\omega)\} \leqslant \rho/\omega$$

since ω and $(h-d)$ are both positive, while the equality conditions, for $y(0)$ and $y(\tau)$, give two equations from which two independent parameters, the angular frequency ω and the 'mark-space' ratio τ/T, can be found. For the 'symmetrical' case where $d = -h$, if $r = (a+b)/2$ or $G(0)$ is infinite, we can take $\tau = T/2$ and the conditions then reduce to

$$\text{Im}\Lambda_0(0,\omega) = -\frac{(a-b)}{2h}, \qquad \text{Re}\Lambda_0(0,\omega) \leqslant \frac{\rho}{\omega},$$

on defining

$$\Lambda_0(\theta,\omega) = \Lambda(\theta,\omega) - \Lambda(\pi+\theta,\omega).$$

When the relay has three states instead of two, the same principles can still be applied, although the details become more involved. For simplicity, we consider only the odd-symmetric characteristic shown in Fig. 4.17, and take $r = 0$, though it is worth noting that, if $G(s)$ has a pole at $s = 0$,

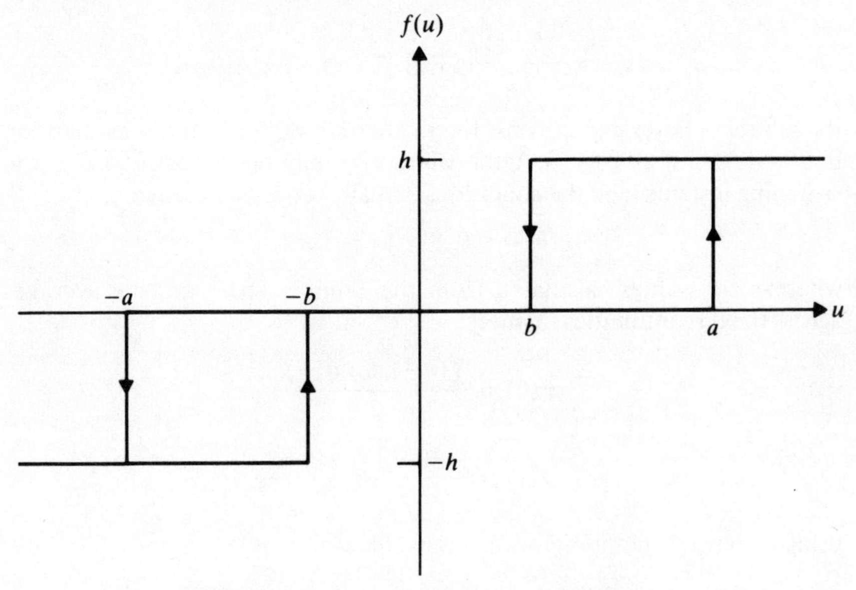

Fig. 4.17 Relay with dead zone and hysteresis.

the value of r is irrelevant, provided it is constant. The relay output may then be assumed to have the form

$$w(t) = h\{v(t) - v(t-T/2)\}$$

as illustrated in Fig. 4.18, $v(t)$ being as before, with $\tau < T/2$, so that $w(t)$ has the Fourier expansion

$$w(t) = \frac{h}{2\pi i}\sum_{k=-\infty}^{\infty}\frac{\{1-(-1)^k\}\{1-\exp(-ik\omega\tau)\}}{k}\exp(ik\omega t)$$

since $\omega T = 2\pi$. Applying the switching conditions at $t = 0$ and $t = \tau$, we now obtain

$$\mathrm{Im}\{\Lambda_0(0,\omega) - \Lambda_0(\theta,\omega)\} = -2a/h$$

$$\mathrm{Im}\{\Lambda_0(0,\omega) - \Lambda_0(-\theta,\omega)\} = 2b/h$$

$$\mathrm{Re}\{\Lambda_0(0,\omega) - \Lambda_0(\theta,\omega)\} \leqslant \rho/\omega$$

$$\mathrm{Re}\{\Lambda_0(0,\omega) - \Lambda_0(-\theta,\omega)\} \leqslant \rho/\omega$$

where $\theta = \omega\tau = 2\pi\tau/T$. From the assumed symmetry properties, it follows that there is no need to consider any other switching times, except by way of assuring ourselves that no spurious ones exist. This can be done by computing the predicted relay input signal $u(t)$, and checking that it does not pass through the switching points at any instants other than those

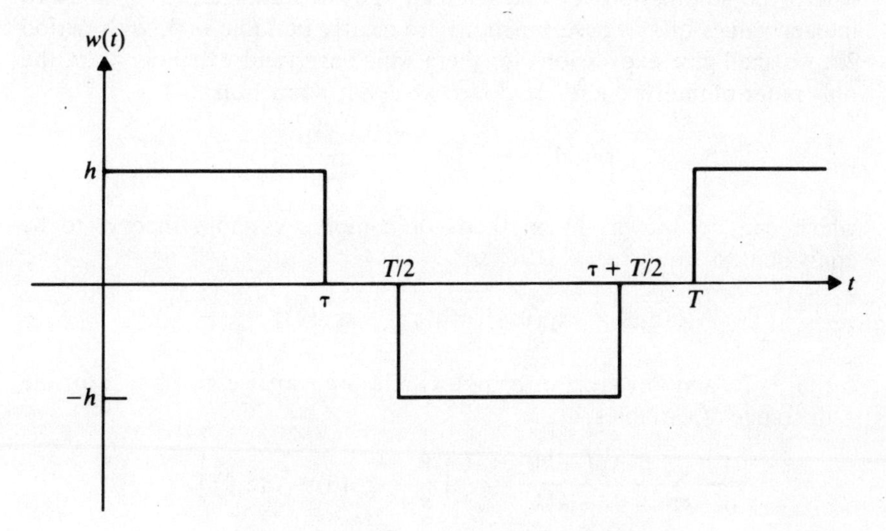

Fig. 4.18 Pulse train with alternating signs.

envisaged. If it does so, then the solution is invalid, which may indicate that the form assumed for the oscillation was not correct; these possibilities, in fact, are always present in applications of Tsypkin's method and constitute its main weaknesses, since the analysis is otherwise exact. A further disadvantage is that there is no simple way to determine the stability of the solutions, in general, though one can always do this by resorting to the time domain, if necessary. It should also be noted that there is nothing, in the foundations of the method, which restricts us to the simple forms of oscillation considered here; any kind of periodic motion may be similarly analysed, and it is indeed possible for systems of this type to possess arbitrarily complicated limit cycles, including unsymmetrical solutions in symmetrical cases, and even chaotic behaviour as in Example 4.2, although this is beyond the scope of the Tsypkin approach.

4.4 Calculation of Tsypkin functions

In order to use Tsypkin's method, it is necessary to be able to compute the functions $\Lambda(\theta,\omega)$ defined above. We note that, since $G(-ik\omega)$ is always the complex conjugate of $G(ik\omega)$, the definitions can be rewritten as

$$\mathrm{Re}\Lambda(\theta,\omega) = \frac{1}{\pi}\sum_{k\neq 0} G(ik\omega)\exp(-ik\theta)$$

$$\mathrm{Im}\Lambda(\theta,\omega) = \frac{1}{i\pi}\sum_{k\neq 0} \frac{G(ik\omega)\exp(-ik\theta)}{k}$$

where the summations extend over all (positive and negative) nonzero integer values of k. These functions are clearly periodic in θ, with period 2π; we shall give expressions for them which are valid when $|\theta| < 2\pi$, the only range of interest here. To do so, we define a function

$$\psi(z,\theta) = \frac{1}{\pi}\sum_{k\neq 0} \frac{\exp(-ik\theta)}{z+ik}$$

which can be shown, by methods of complex variable theory, to be equivalent to

$$\psi(z,\theta) = \exp(\theta z)\{\coth(\pi z) - \mathrm{sgn}(\theta)\} - \frac{1}{\pi z}$$

for $|\theta| < 2\pi$ and any (real or complex) z; moreover, we also have, for the same range of variables,

$$\frac{1}{i\pi}\sum_{k\neq 0} \frac{\exp(-ik\theta)}{(z+ik)k} = \frac{1}{z}\left(\frac{\theta}{\pi} - \mathrm{sgn}(\theta) - \psi(z,\theta)\right).$$

Hence, if $G(s)$ has a partial fraction expansion

$$G(s) = \sum_{j=1}^{n} \frac{r_j}{s+c_j}$$

involving only simple poles, it follows that

$$\mathrm{Re}\Lambda(\theta,\omega) = \frac{1}{\omega}\sum_{j=1}^{n} r_j\psi\left(\frac{c_j}{\omega}, \theta\right)$$

$$\mathrm{Im}\Lambda(\theta,\omega) = \sum_{j=1}^{n} \frac{r_j}{c_j}\left\{\frac{\theta}{\pi} - \mathrm{sgn}(\theta) - \psi\left(\frac{c_j}{\omega}, \theta\right)\right\}$$

and similar expressions, for cases with multiple poles, can easily be derived by differentiation with respect to the pole positions. It may also be noted that the effect of attaching a time-delay factor, $\exp(-\lambda s)$, to $G(s)$, would be simply to replace θ by $(\theta+\lambda\omega)$ in these formulae. Of particular importance for odd-symmetric cases are the functions $\Lambda_0(\theta,\omega)$, which may be obtained by omitting the terms with even k and doubling those with odd k, so that $\Lambda_0(\pi+\theta,\omega) = -\Lambda_0(\theta,\omega)$; if $G(s)$ is a rational function with simple poles as above, we find

$$\mathrm{Re}\Lambda_0(\theta,\omega) = \frac{1}{\omega}\sum_{j=1}^{n} r_j\,\psi_0\left(\frac{c_j}{\omega},\theta\right)$$

$$\mathrm{Im}\Lambda_0(\theta,\omega) = -\sum_{j=1}^{n}\frac{r_j}{c_j}\left(\mathrm{sgn}(\theta) + \psi_0\left(\frac{c_j}{\omega},\theta\right)\right)\}$$

where

$$\psi_0(z,\theta) = \exp(\theta z)\,\{\tanh(\pi z/2) - \mathrm{sgn}(\theta)\}$$

for $|\theta| < \pi$. Graphs of the functions Λ, Λ_0 and combinations thereof, usually plotted against ω for fixed values of θ, are sometimes known as Tsypkin loci.

Example 4.5 System containing a pure integrator

If $G(0)$ is infinite, the limit cycle equations for the system described by Figs. 3.12 and 4.16 yield

$$\frac{\tau}{T} = \frac{-d}{h-d}$$

assuming that $d < 0 < h$, and

$$\mathrm{Im}\left(\frac{\Lambda(\theta,\omega) + \Lambda(-\theta,\omega)}{2} - \Lambda(0,\omega)\right) = \frac{a-b}{h-d}$$

while the inequality conditions become

$$\mathrm{Re}\{\Lambda(0,\omega) - \Lambda(\pm\theta,\omega)\} \leqslant \rho/\omega$$

with $\theta = \omega\tau$. Because of the pole in $G(s)$ at $s = 0$, we have to consider the limiting behaviour of $\psi(z,\theta)$ as $z \to 0$, in order to calculate the Tsypkin functions by using the formulae derived above. For the simple case of a pure integrator, with

$$G(s) = 1/s$$

this procedure gives

$$\mathrm{Re}\Lambda(\theta,\omega) = \frac{1}{\omega}\left(\frac{\theta}{\pi} - \mathrm{sgn}(\theta)\right)$$

$$\mathrm{Im}\Lambda(\theta,\omega) = \frac{-1}{\omega}\left(\frac{\pi}{3} - |\theta| + \frac{\theta^2}{2\pi}\right)$$

for $|\theta| < 2\pi$. Since $\rho = 1$ in this case, the inequality constraints are automatically satisfied, and a limit cycle exists, with period $T = 2\pi/\omega$, given by

$$T = \left(\frac{a-b}{h-d}\right)\left\{\left(1-\frac{\theta}{2\pi}\right)\frac{\theta}{2\pi}\right\}^{-1} = (a-b)\ \left(\frac{1}{h}-\frac{1}{d}\right)$$

snce $\theta = 2\pi\tau/T$.

Example 4.6 A comparison with the describing function method

In the system of Example 3.5, the relay element has the characteristic corresponding to Fig. 4.16, with $h = -d = 1$ and $b = -a$, so that the Tsypkin conditions for a limit cycle become

$$\text{Im}\Lambda_0(0,\omega) = -a$$

$$\text{Re}\Lambda_0(0,\omega) \leqslant 0$$

since $\rho = 0$. The transfer function $G(s)$ is given by

$$G(s) = \frac{K}{s} - \frac{K}{s+1}$$

whence we obtain

$$\text{Re}\Lambda_0(0,\omega) = -\frac{K}{\omega}\ \tanh\left(\frac{\pi}{2\omega}\right)$$

$$\text{Im}\Lambda_0(0,\omega) = K\left\{\tanh\left(\frac{\pi}{2\omega}\right) - \frac{\pi}{2\omega}\right\}$$

by use of the formulae involving $\psi_0(z,\theta)$, derived previously. It follows that both the required conditons are satisfied and therefore a limit cycle occurs, with ω given by

$$\frac{\pi}{2\omega} - \tanh\left(\frac{\pi}{2\omega}\right) = \frac{a}{K}$$

to be compared with the describing function prediction of

$$\frac{4}{\pi\omega(1+\omega^2)} = \frac{a}{K}$$

from Example 3.5. Numerically, the results agree to within about 0.5% in the limit $(a/K)\rightarrow 0$, but the discrepancy becomes considerably larger (around 20%) as $(a/K)\rightarrow\infty$.

Example 4.7 Resonant system with an ideal relay

The so-called ideal relay characteristic is obtained from Fig. 4.16 by setting $d = -h$ and $a = b = 0$, so that it has odd symmetry and no hysteresis. With zero reference input, and transfer function

$$G(s) = \frac{-s}{s^2+2\zeta s+1}$$

where $0<\zeta<1$, in Fig. 3.12, the conditions for a limit cycle, obtained from Tsypkin's method, are

$$\mathrm{Im}\Lambda_0(0,\omega) = 0$$
$$\mathrm{Re}\Lambda_0(0,\omega) \leqslant -1/\omega$$

since $\rho = -1$. Writing $\gamma = \surd(1-\zeta^2)$, so that

$$\mathrm{G}(s) = \frac{1}{2\gamma}\left(\frac{\mathrm{i}(\zeta+\mathrm{i}\gamma)}{s+\zeta+\mathrm{i}\gamma} - \frac{\mathrm{i}(\zeta-\mathrm{i}\gamma)}{s+\zeta-\mathrm{i}\gamma}\right)$$

the required Tsypkin functions turn out to be given by

$$\mathrm{Re}\Lambda_0(0,\omega) = \frac{-1}{\omega\gamma}\left(\frac{\gamma\sinh(\pi\zeta/\omega) + \zeta\sin(\pi\gamma/\omega)}{\cosh(\pi\zeta/\omega) + \cos(\pi\gamma/\omega)}\right)$$

$$\mathrm{Im}\Lambda_0(0,\omega) = \frac{\sin(\pi\gamma/\omega)}{\gamma\{\cosh(\pi\zeta/\omega) + \cos(\pi\gamma/\omega)\}}$$

after some manipulation of complex hyperbolic and trigonometric expressions. Thus, the equality condition is satisfied by $\omega = \gamma/k$ for any positive integer k, but the inequality then restricts us to odd values of k. Moreover, a check on the form of the relay input reveals that the only valid solution is obtained with $k = 1$, giving a limit cycle of period $2\pi/\gamma$.

4.5 Relay systems with periodic inputs

Tsypkin's method can also be applied in some cases where the external reference input to the system is a periodic function of time, rather than a constant. It is then assumed that the relay output is periodic, with the same frequency as the reference signal, or a submultiple thereof; this assumption thus effectively excludes the consideration of an internally generated oscillation being present simultaneously, though it allows for the possible occurrencc of subharmonics. The analysis follows the same lines as for the autonomous case, except in two respects: first, the period of oscillation is now specified, instead of being determined by the switching conditions;

second, there is an unknown phase difference between the external and internal signals. Application of the switching conditions for the relay then gives the same number of equations as before, with the same number of parameters to be determined, since the unknown phase in the forced case replaces the unknown frequency in the case of a free system.

4.6 Exercises

1 Set up a state-space representation for Example 4.6, and show that the point transformation method gives the same results as the Tsypkin conditions for a limit cycle.

2 Apply the point transformation method to a state-space form of Example 4.7. Show how the results compare with those of Tsypkin's method and with a describing function analysis.

3 Analyse the behaviour of the variable-structure system governed by the equations

$$\begin{aligned} \dot{x}_1 &= x_2 \\ \dot{x}_2 &= -x_1 + \{\mathrm{sgn}(\sigma x_1) - 1\}\, x_2 \end{aligned}$$

where $\sigma = \lambda x_1 + x_2$ with $\lambda > 1$, and make a sketch of the phase portrait.

4 Apply the Tsypkin limit cycle conditions to Example 4.1, and relate the results to those obtained from the point transformation method.

5 Compare the results of Tsypkin's method with the describing function predictions for a system as in Example 4.6, with zero reference input and the transfer function altered to $G(s) = 1/(s+1)$.

6 Replacing the transfer function in Example 4.7 by

$$G(s) = \frac{\exp(-s)}{s}$$

show that infinitely many limit cycles can occur and find their periods.

5 STABILITY

Although the concept of stability has been repeatedly used in earlier chapters, we have not yet subjected it to detailed examination, other than in a purely local context. That is to say, when determining whether or not a particular motion or equilibrium of a system should be regarded as stable, we have considered only infinitesimal perturbations around the nominal solution. This procedure is indeed adequate for linear systems, where the local and global properties are identical, but ceases to be so when nonlinearities are introduced, since the effect of a disturbance input, or a change of initial conditions, may then be crucially dependent on its magnitude. The subject in fact becomes rather complicated, because many different types of stability can be defined, according to the boundedness and convergence properties satisfied; moreover, they can be defined in either of the main analytical frameworks available, namely the state-space and the input-output relation, but the connection between these two structures is no longer simple. In particular, with regard to the (nonlocal) stability properties of nonstatic equilibrium solutions, such as limit cycles, it is doubtful that the input–output description has any significant part to play at all; it does, however, provide an important alternative approach to the derivation of global stability criteria, as will be seen, especially in the context of feedback systems.

We will, in any case, begin with the state-space formulation, concentrating our attention at first on the stability of autonomous systems and their static equilibria. One can actually reduce the study of stability for an arbitrary trajectory to the case of an equilibrium point, by a change of variables, but the resulting equations are, in general, time-dependent and consequently much more difficult to handle, even when linearised. The stability analysis of limit cycles thus properly belongs in the theory of non-autonomous systems, which will be deferred to a later stage.

5.1 Equilibria of autonomous systems

Suppose that an autonomous system, described by the vector differential equation

$$\dot{\mathbf{x}} = \mathbf{f}(\mathbf{x})$$

has an equilibrium point at $\hat{\mathbf{x}}$, so that

$$\mathbf{f}(\hat{\mathbf{x}}) = 0$$

The local stability properties of the equilibrium can be obtained by linearisation, but in order to go beyond this, we need to introduce some more precise definitions, as follows.

We shall say that the equilibrium is *stable* if, given any $\epsilon > 0$, there exists $\delta > 0$ such that $|\mathbf{x}(0)-\hat{\mathbf{x}}|<\delta$ implies $|\mathbf{x}(t)-\hat{\mathbf{x}}|<\epsilon$ for all $t > 0$; that is to say, a trajectory can be made to remain within an arbitrarily small distance from $\hat{\mathbf{x}}$ by starting sufficiently close to it. Also, we define the equilibrium to be *convergent* if there exists $\Delta > 0$ such that $|\mathbf{x}(0)-\hat{\mathbf{x}}|<\Delta$ implies that $\mathbf{x}(t)\to\hat{\mathbf{x}}$ as $t\to\infty$, which means that all trajectories starting near enough to $\hat{\mathbf{x}}$ approach it asymptotically. Further, we call the equilibrium *asymptotically stable* if it is both stable and convergent. It should be noted that stability and convergence are independent properties; the fact that stability does not imply convergence is obvious, but it is also possible to find examples, admittedly somewhat artificial, which show that an equilibrium can be both convergent and unstable.

The properties defined above are nonlocal, in the sense that they are not based on infinitesimal deviations from equilibrium, but they still do not allow for such deviations being arbitrarily large. To cover this possibility, we say that the equilibrium is *globally asymptotically stable* if it is stable and all trajectories converge to $\hat{\mathbf{x}}$ as $t\to\infty$; this evidently implies that the system can have no other equilibrium points, and also that every trajectory remains bounded for all $t>0$. On the other hand, if an equilibrium point is asymptotically stable but not globally so, we wish to quantify the extent of its convergence in some way. For this purpose, it is convenient to utilise the concept of an 'invariant set', which is defined as any set of points in the state-space, such that each trajectory starting within it remains there for all subsequent times; it is thus carried into itself by the evolution of the system and is, in fact, just a collection of the 'future parts' of trajectories. We then define a 'domain of attraction', about an asymptotically stable equilibrium point $\hat{\mathbf{x}}$, as an open invariant set containing $\hat{\mathbf{x}}$, with the property that every trajectory which starts in it approaches $\hat{\mathbf{x}}$ as $t\to\infty$. The requirement that the set is open, which means that every point in it has a neighbourhood entirely contained within it, is introduced in order to prevent the definition being trivially satisfied, for instance, by a single point or trajectory. Further, we define the 'maximal domain of attraction' as the set of all points from which trajectories converge to the equilibrium, or equivalently, as the union of all domains of attraction about $\hat{\mathbf{x}}$. It follows immediately that, if the equilibrium is globally asymptotically stable, then the maximal domain of attraction is the entire state-space.

5.2 Lyapunov's methods

The investigation of local stability properties by linearisation is sometimes known as the first method of Lyapunov, but his name is more commonly associated with his second (or direct) method, which involves the utilisation of certain auxiliary functions, called Lyapunov functions. These are simply continuous scalar functions of the state variables, with continuous partial derivatives, so that the time-derivative of such a function $V(\mathbf{x})$ along a trajectory, given by

$$\dot{V}(\mathbf{x}) = \sum_{j=1}^{n} \frac{\partial V}{\partial x_j} \dot{x}_j = \sum_{j=1}^{n} \frac{\partial V}{\partial x_j} f_j(\mathbf{x})$$

is itself a continuous function of $\mathbf{x}$; further, both V and $\dot{V}$ are required to satisfy certain constraints, from which information about stability can be deduced.

To be more specific, suppose there is an open connected set, denoted by S, containing an equilibrium point $\hat{\mathbf{x}}$, such that, for all $\mathbf{x}$ in S, the following properties hold:

(i) $V(\mathbf{x}) > V(\hat{\mathbf{x}})$ whenever $\mathbf{x} \neq \hat{\mathbf{x}}$;
(ii) if S is unbounded, then $V(\mathbf{x}) \to \infty$ as $|\mathbf{x}| \to \infty$.

Property (i) is sometimes known as positivity (or positive-definiteness) since we can always make $V(\hat{\mathbf{x}}) = 0$ by adding an appropriate constant; property (ii) is called radial unboundedness and becomes important if, for example, S is the whole state-space, but is not necessary if S is bounded. Now, since S is open, it contains a closed ball $|\mathbf{x}-\hat{\mathbf{x}}| \leqslant R$, for some $R > 0$, and we can define the functions

$$\psi(\delta) = \sup_{|\mathbf{x}-\hat{\mathbf{x}}|\leqslant\delta} \{V(\mathbf{x})-V(\hat{\mathbf{x}})\}$$

$$\chi(\delta) = \inf_{\delta\leqslant|\mathbf{x}-\hat{\mathbf{x}}|\leqslant R} \{V(\mathbf{x})-V(\hat{\mathbf{x}})\}$$

for all $\delta \leqslant R$. From the continuity of $V(\mathbf{x})$, it follows that both $\psi(\delta)$ and $\chi(\delta)$ are continuous nondecreasing functions of δ, satisfying

$$\psi(0) = \chi(0) = 0$$

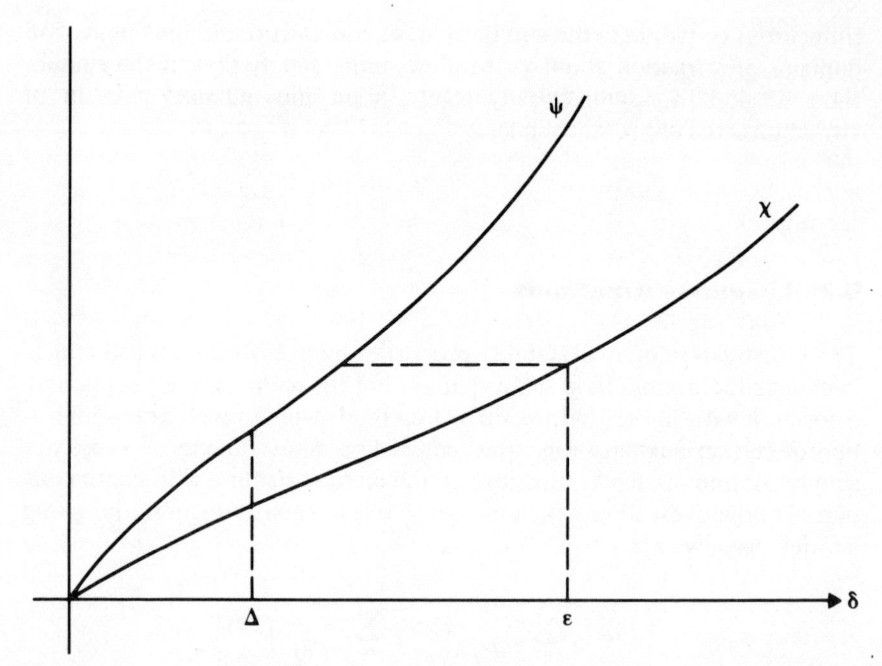

Fig. 5.1 The functions $\psi(\delta)$ and $\chi(\delta)$.

and also, for $\delta > 0$,

$$\psi(\delta) \geqslant \chi(\delta) > 0$$

because of (i); the relationship between these functions is illustrated in Fig. 5.1.

We can now consider the implications of various conditions which may be satisfied by $\dot{V}$. To begin with, suppose we have

(iii) $\dot{V}(\mathbf{x}) \leqslant 0$ for all $\mathbf{x}$ in S,

so that $V(\mathbf{x})$ is non-increasing with respect to time along any trajectory, so long as it remains within S. Then, given any ϵ satisfying $0 < \epsilon \leqslant R$, we can choose Δ such that

$$0 < \psi(\Delta) < \chi(\epsilon)$$

and hence, if $|\mathbf{x}(0) - \hat{\mathbf{x}}| \leqslant \Delta$, it follows immediately that $|\mathbf{x}(t) - \hat{\mathbf{x}}| < \epsilon$ for all $t \geqslant 0$, which means that the equilibrium at $\hat{\mathbf{x}}$ is stable. Next, together with (iii), let us impose the further condition

(iv) $\dot{V}(\mathbf{x})$ does not vanish identically along any trajectory in S other than $\mathbf{x}(t) \equiv \hat{\mathbf{x}}$.

Now, since any trajectory which starts sufficiently near $\hat{\mathbf{x}}$ remains bounded, it must asymptotically approach its positive limit set Ω as $t \to \infty$. Also, since $\{V(\mathbf{x}) - V(\hat{\mathbf{x}})\}$ is non-increasing with t and bounded below by zero, it tends to a non-negative limit as $t \to \infty$, along any such trajectory which stays in S, and this limit must therefore be its value at every point in Ω. However, it follows from the definition of Ω that it is an invariant set and thus contains at least one trajectory, on which V must be constant and hence $\dot{V} = 0$, so that by (iv), it consists only of the point $\hat{\mathbf{x}}$. Consequently, for any $\mathbf{x}$ in Ω, $V(\mathbf{x}) = V(\hat{\mathbf{x}})$, and therefore Ω is just the equilibrium point itself, so we have shown that every trajectory with $|\mathbf{x}(0) - \hat{\mathbf{x}}| \leq \Delta$, for Δ chosen as above, approaches $\hat{\mathbf{x}}$ asymptotically as $t \to \infty$. The equilibrium is thus asymptotically stable, as a consequence of conditions (iii) and (iv); moreover, we may note that, in order for these conditions to be satisfied, it is sufficient that $\dot{V}(\mathbf{x}) < 0$ for all $\mathbf{x}(\neq \hat{\mathbf{x}})$ in S.

In the foregoing analysis, we have not made any use of property (ii), that is to say, the assumption that V is radially unbounded. If, however, S is the entire state-space, this property enables us to take R infinite in the definitions of $\psi(\delta)$ and $\chi(\delta)$, with the result that both these functions tend to infinity as $\delta \to \infty$. It then follows, by the preceding arguments, that condition (iii) implies that every trajectory is bounded; further, if condition (iv) is also satisfied, then all trajectories approach $\hat{\mathbf{x}}$ as $t \to \infty$, so that the equilibrium is globally asymptotically stable. Actually, in order to establish the boundedness of all trajectories, which is sometimes called 'Lagrange stability', we do not need property (i), or even the assumption that an equilibrium point exists; it is sufficient that (ii) and (iii) should hold for a region S which is the exterior of some bounded set, on whose boundary $V(\mathbf{x})$ has a finite supremum.

One of the main uses of Lyapunov functions is in the estimation of domains of attraction. In fact, the procedure we have just used, in the proof of asymptotic stability, provides a means of doing this, but it is usually more convenient to work with regions specifically related to the Lyapunov function, as follows. If we have an open region S, consisting of points where $V(\mathbf{x}) < c$, for some constant c, with the boundary of S given by $V(\mathbf{x}) = c$, such that condition (iii) holds, then trajectories starting in S cannot leave it and, if (iv) also holds, then all such trajectories converge to the equilibrium point $\hat{\mathbf{x}}$, so that S is a domain of attraction. We would clearly wish to choose c as large as possible, and in some, though not all, cases this can be done by setting it equal to the minimum value of $V(\mathbf{x})$ over all points (apart from $\hat{\mathbf{x}}$) where $\dot{V}(\mathbf{x}) = 0$; for this to work, it is necessary that $V(\mathbf{x}) < 0$ for all $\mathbf{x}$ sufficiently near $\hat{\mathbf{x}}$. Further, the procedure can be generalised to a wider context by taking S to consist of points satisfying $b < V(\mathbf{x}) < c$, where b and c are constants, so that the boundary of the region has two distinct components given by $V(\mathbf{x}) = b$ and $V(\mathbf{x}) = c$. Then, if conditions (iii) and (iv) hold, with the reference to $\hat{\mathbf{x}}$ in (iv) removed, it follows that every trajectory which starts in S must ultimately leave it, by entering the region $V(\mathbf{x}) \leq b$, from which it cannot return and which therefore contains its positive limit set. Thus, as illustrated in Fig. 5.2, all

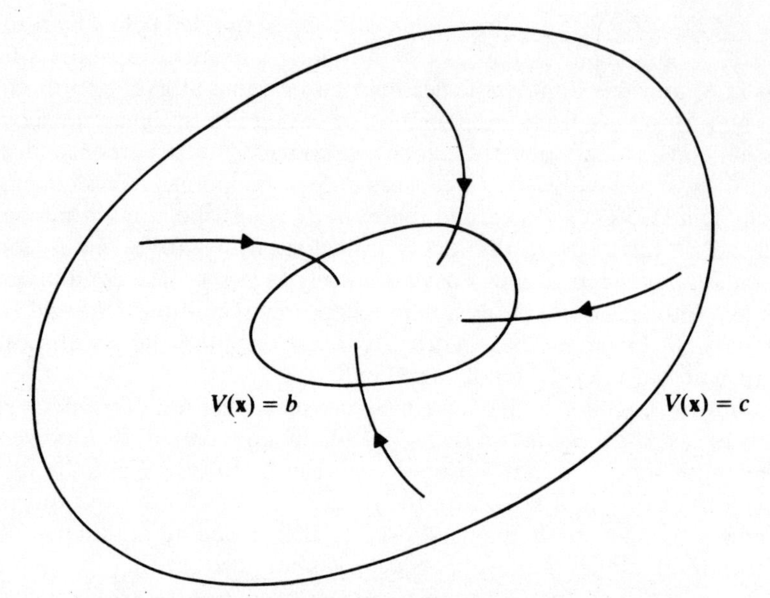

Fig. 5.2 Region bounded by hypersurfaces of constant V.

trajectories originating in a certain region are eventually confined within a smaller one, which is a subset of the first. Moreover, if the stronger condition $\dot{V}(\mathbf{x}) \leqslant K\{b - V(\mathbf{x})\}$ is satisfied everywhere in S, for some positive constant K, then we can estimate the rate of convergence, since $\{V(\mathbf{x}) - b\}\exp(Kt)$ remains finite as $t \to \infty$.

5.3 Construction of Lyapunov functions

The original motive for the development of Lyapunov's direct method was based on the physical concept of the energy content of a system, which, in the usual dissipative case, is naturally a decreasing function of time, and this is often a fruitful source of Lyapunov functions in practice. For example, the function $V(\mathbf{x})$ used in Example 2.1 can be derived in this way. On the other hand, there is no reason why we should be restricted to using functions of this type, and indeed it may not be appropriate in many cases. There is, unfortunately, no completely general systematic procedure for obtaining Lyapunov functions, but there are several useful approaches. One method is simply to take a function, based on energetic or other considerations, containing some free parameters, calculate its time-derivative, and then adjust the parameters until V and $\dot{V}$ satisfy the required conditions, if this is possible. Also, if the system, when linearised about an equilibrium point $\hat{\mathbf{x}}$, is asymptotically stable, we can obtain an infinite number of Lyapunov functions, valid at least in some neighbourhood of $\hat{\mathbf{x}}$, from the linearisation, as follows. Writing the linearised equations as

$$\dot{\tilde{\mathbf{x}}} = \mathbf{A}\tilde{\mathbf{x}}$$

where $\tilde{\mathbf{x}} = \mathbf{x} - \hat{\mathbf{x}}$, the necessary and sufficient condition for asymptotic stability is that all the eigenvalues of the constant matrix $\mathbf{A}$ should have negative real parts, so that

$$\exp(\mathbf{A}t) \rightarrow 0$$

as $t \rightarrow \infty$. In that case, for any constant matrix $\mathbf{Q}$, we can construct a matrix $\mathbf{P}$ from the convergent integral

$$\mathbf{P} = \int_0^\infty \exp(\mathbf{A}^{\mathrm{T}}t)\, \mathbf{Q} \exp(\mathbf{A}t)\, \mathrm{d}t$$

which then satisfies

$$\mathbf{PA} + \mathbf{A}^{\mathrm{T}}\mathbf{P} = \int_0^\infty \frac{\mathrm{d}}{\mathrm{d}t} \{\exp(\mathbf{A}^T t)\mathbf{Q} \exp(\mathbf{A}t)\}\, \mathrm{d}t = -\mathbf{Q}.$$

Moreover, if $\mathbf{Q}$ is symmetric, then so is $\mathbf{P}$, by construction, and if $\mathbf{Q}$ is positive-definite, so also is $\mathbf{P}$ since, for any non-vanishing vector $\mathbf{v}$,

$$\mathbf{v}^{\mathrm{T}}\mathbf{P}\mathbf{v} = \int_0^\infty \{\exp(\mathbf{A}t)\mathbf{v}\}^{\mathrm{T}}\, \mathbf{Q}\{\exp(\mathbf{A}t)\mathbf{v}\}\mathrm{d}t > 0.$$

Consequently, given any symmetric positively-definite $\mathbf{Q}$, we can form a function

$$V = \tilde{\mathbf{x}}^{\mathrm{T}}\, \mathbf{P}\tilde{\mathbf{x}}$$

such that

$$\dot{V} = \tilde{\mathbf{x}}^{\mathrm{T}}\, (\mathbf{PA} + \mathbf{A}^{\mathrm{T}}\mathbf{P})\tilde{\mathbf{x}} = -\tilde{\mathbf{x}}^{\mathrm{T}}\mathbf{Q}\tilde{\mathbf{x}}$$

according to the linearised equations, whence $V > 0$ and $\dot{V} < 0$ for all $\tilde{\mathbf{x}} \neq 0$. Further, when the original nonlinear equations are used instead of the linear model, $\dot{V}$ will still have the required properties in some open region around $\hat{\mathbf{x}}$, so that V can be used as a Lyapunov function. It should be noted also, that in order to construct V, we can obtain $\mathbf{P}$ algebraically by solving the 'Lyapunov matrix equation'

$$\mathbf{PA} + \mathbf{A}^{\mathrm{T}}\mathbf{P} = -\mathbf{Q}$$

without needing to evaluate the integral expression.

Another method which deserves mention, if only on theoretical grounds, is that of Zubov. Here, one chooses a function $W(\mathbf{x})$, with the properties desired for $\dot{V}(\mathbf{x})$, namely $W(\hat{\mathbf{x}}) = 0$ and $W(\mathbf{x}) < 0$ for all $\mathbf{x} \neq \hat{\mathbf{x}}$, and then computes $V(\mathbf{x})$ by solving the partial differential equation

$$\sum_{j=1}^{n} \frac{\partial V}{\partial x_j} f_j(\mathbf{x}) = W(\mathbf{x})$$

with the boundary condition $V(\hat{\mathbf{x}}) = 0$. If the equilibrium at $\hat{\mathbf{x}}$ is asymptotically stable, the solution $V(\mathbf{x})$ will be positive for $\mathbf{x} \neq \hat{\mathbf{x}}$ and will tend to infinity on the boundary of the maximal domain of attraction, which can thus be computed, in principle, though there may be severe numerical difficulties in practice.

Example 5.1 An oscillator with nonlinear damping

This is a variant of the Van der Pol system in Example 2.2, with a different kind of nonlinearity in the damping term, such that the state-space equations become

$$\dot{x}_1 = x_2$$
$$\dot{x}_2 = -x_1 - 2\zeta(x_2 - x_2^3 + \lambda x_2^5)$$

for some positive constants ζ and λ. The only equilibrium point is therefore at $x = 0$, and the model obtained by linearisation around it (for $\zeta < 1$) is a damped harmonic oscillator with a damping ratio ζ. However, if λ is small enough, a describing function analysis indicates the existence of two limit cycles, both represented approximately by circles in the phase plane, the outer one being stable and the inner unstable, as shown in Fig. 5.3. It is thus to be expected that trajectories starting inside the unstable limit cycle will converge to the origin, and those starting outside it will approach the stable limit cycle asymptotically. A system of this type, where sustained oscillatory behaviour only occurs if the initial point is outside some domain of attraction about a stable equilibrium, is sometimes called a 'hard' oscillator, as distinct from a self-exciting or 'soft' one, such as is described by the usual Van der Pol equation. To estimate the domain of attraction by Lyapunov's method, the obvious function to choose is the 'energy' expression

$$V(\mathbf{x}) = x_1^2 + x_2^2$$

so that

$$\dot{V}(\mathbf{x}) = -4\zeta(1 - x_2^2 + \lambda x_2^4)\, x_2^2$$

from the state equations. It is clear that $\dot{V}$ cannot vanish identically along any nontrivial trajectory, since such a trajectory would have $x_2 =$ constant $\neq 0$ and hence $x_1 =$ constant but $\dot{x}_1 \neq 0$, giving a contradiction. Also, for $|x_2|$ sufficiently small, $\dot{V} \leqslant 0$ and so the interior of any contour of V lying in this region is a domain of attraction, the contours being, of course, circles centred at the origin. In fact, if $\lambda \geqslant 1/4$, we have $\dot{V} \leqslant 0$ everywhere and thus, since V is radially unbounded, the system is globally asymptotically

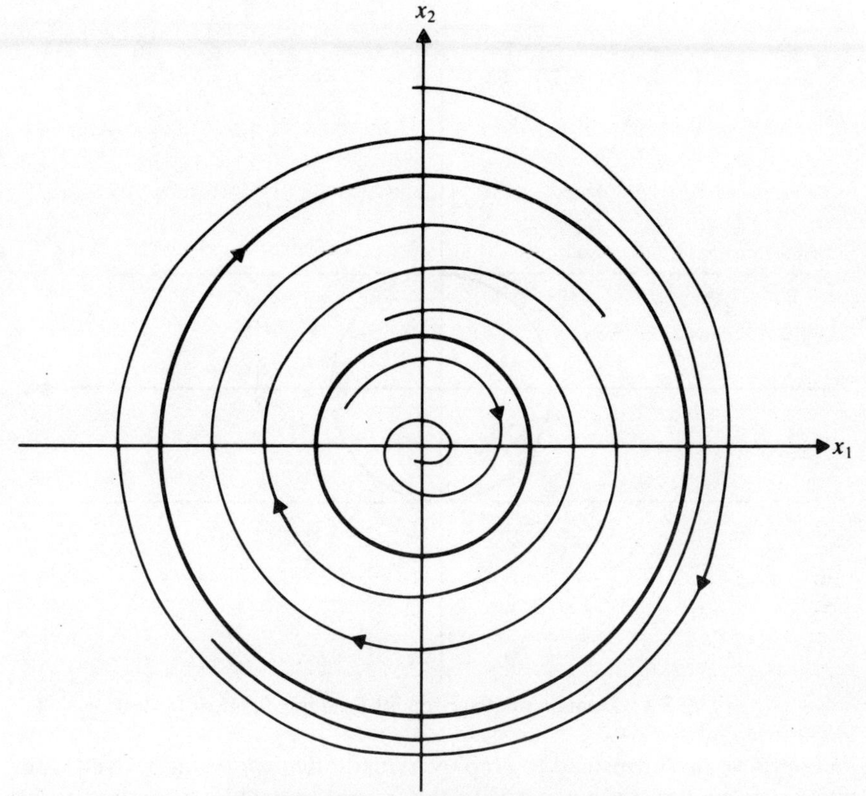

Fig. 5.3 Stable and unstable limit cycles.

stable, so no limit cycle occurs. For $\lambda < 1/4$, a domain of attraction is given by

$$x_1^2 + x_2^2 < \frac{1 - \sqrt{(1 - 4\lambda)}}{2\lambda}$$

as in Fig. 5.4.

Example 5.2 Use of the Lyapunov matrix equation

To illustrate this technique, let us consider a damped nonlinear oscillator, described by the differential equation

$$\ddot{y} + 2\zeta\dot{y} + (1-y)y = 0$$

where ζ is constant, with $0 < \zeta < 1$. It would be possible to construct a Lyapunov function representing the energy of the system, using the fact that

$$\frac{\mathrm{d}}{\mathrm{d}t}(\dot{y}^2 + y^2 - \tfrac{2}{3}y^3) = -4\zeta\dot{y}^2$$

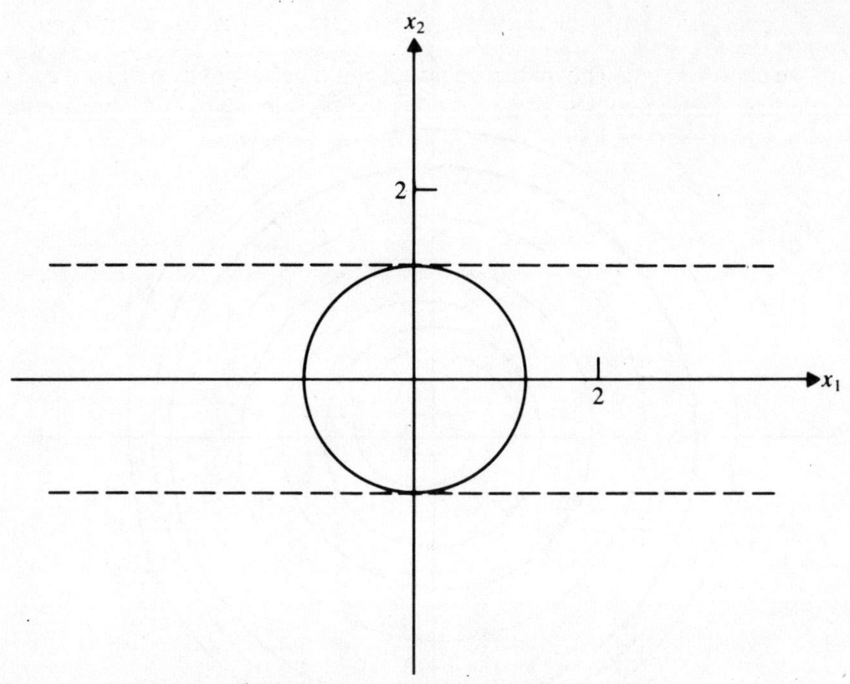

Fig. 5.4 Domain of attraction for Example 5.1 ($\lambda = 0.2$).

but here we prefer instead to employ a quadratic expression derived from the Lyapunov matrix equation. In this case, the analysis turns out to be considerably simplified if we use the state-space representation obtained by setting

$$x_1 = y$$

$$x_2 = \frac{\dot{y} + \zeta y}{\gamma}$$

where $\gamma = \sqrt{(1-\zeta^2)}$, so that the state equations are

$$\dot{x}_1 = \quad -\zeta x_1 + \gamma x_2$$

$$\dot{x}_2 = -\gamma x_1 - \zeta x_2 + \frac{x_1^2}{\gamma}$$

and hence there are two equilibrium points, at (0,0) and $(1,\zeta/\gamma)$. These are, respectively, a stable focus and a saddle point; the plant matrix of the system linearised around the origin is clearly

$$\mathbf{A} = \begin{pmatrix} -\zeta & \gamma \\ -\gamma & -\zeta \end{pmatrix}$$

which is to be inserted into the Lyapunov matrix equation, namely

$$\mathbf{PA} + \mathbf{A}^T\mathbf{P} = -\mathbf{I}$$

if we choose $\mathbf{Q} = \mathbf{I}$. The matrix equation could be solved by writing it out as a set of simultaneous scalar equations for the elements of **P**, or in many other ways, but the solution here is evidently given by

$$\mathbf{P} = \mathbf{I}/2\zeta$$

whence

$$V(x) = \frac{x_1^2 + x_2^2}{2\zeta}$$

and so

$$\dot{V}(x) = -(x_1^2 + x_2^2) + \frac{x_1^2 x_2}{\zeta\gamma}$$

from the state-space equations. If we now set $\dot{V} = 0$, solve for x_1 and substitute into $V(x)$, we obtain

$$V = \frac{x_2^3}{2\zeta(x_2 - \zeta\gamma)}$$

on the curve $\dot{V}(x) = 0$, which exists only for $x_2 > \zeta\gamma$. By elementary calculus, the minimum of V on this curve is found to occur at $x_2 = 3\zeta\gamma/2$, where $V = 27\zeta\gamma^2/8$, and consequently a domain of attraction

$$x_1^2 + x_2^2 < 27\zeta^2\gamma^2/4$$

as illustrated in Fig. 5.5, is obtained.

5.4 Unstable equilibrium points

If an equilibrium solution fails to be stable, it is not usually of much interest for practical purposes, and so, although instability criteria can be derived, along the lines of Lyapunov's direct method, they are seldom used. Nevertheless, unstable equilibria, whether static or dynamic, may sometimes be of significance, in that they are often associated with boundaries between regions where different types of behaviour are exhibited. Thus, in Example 5.1, an unstable limit cycle is the boundary of the maximal domain of attraction; similarly, the saddle point in Example 5.1, which of course cannot be inside any domain of attraction, may be shown to lie on the boundary of the maximal one, by using the system energy as a Lyapunov function. It is, in fact, quite common for an equilibrium point of type 1 (as defined in Chapter 2), that is to say, where the plant matrix of the linearised system has only one right half-plane eigenvalue, to be

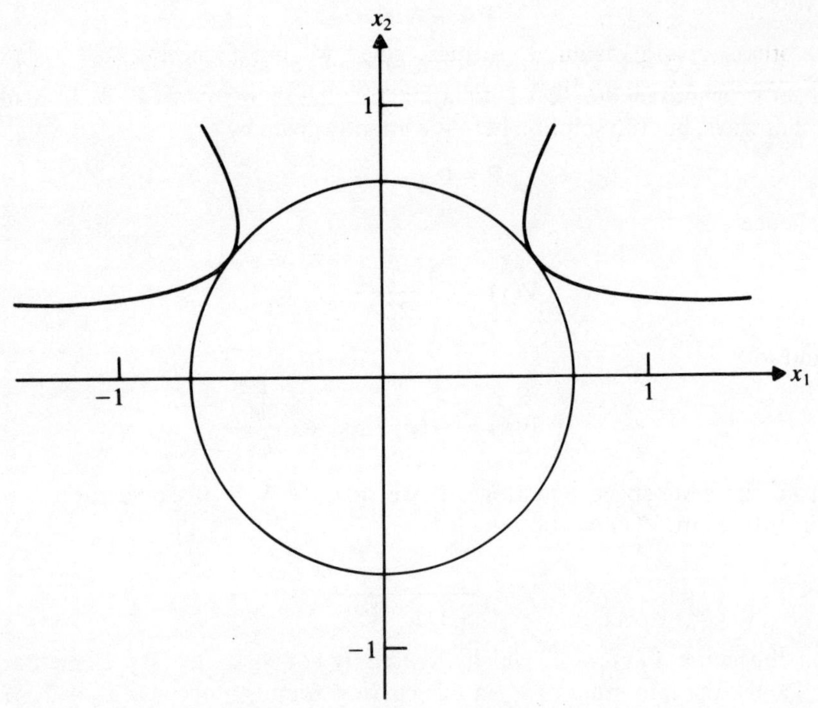

Fig. 5.5 Domain of attraction for Example 5.2 ($\zeta = 0.3$).

located on the boundary of a domain of attraction, which can thus be locally approximated by using the linear model.

Generally, the set of all trajectories which approach an equilibrium asymptotically as $t \to \infty$ is called its 'stable manifold', while those which converge to it as $t \to -\infty$ constitute the 'unstable manifold'. For a type 1 equilibrium point $\hat{\mathbf{x}}$, whose associated plant matrix $\mathbf{A}$ has a positive eigenvalue λ, the stable manifold of the locally linearised model is the hyperplane

$$\mathbf{v}^T(\mathbf{x} - \hat{\mathbf{x}}) = 0$$

where $\mathbf{v}$ is the eigenvector of $\mathbf{A}^T$ corresponding to λ, so that

$$\mathbf{A}^T\mathbf{v} = \lambda\mathbf{v}, \qquad \mathbf{v}^T\mathbf{A} = \lambda\mathbf{v}^T$$

and hence

$$\frac{d}{dt}\{\mathbf{v}^T(\mathbf{x}-\hat{\mathbf{x}})\} = \mathbf{v}^T\dot{\mathbf{x}} = \lambda\mathbf{v}^T(\mathbf{x}-\hat{\mathbf{x}})$$

from the linearised equations

$$\dot{\mathbf{x}} = \mathbf{A}(\mathbf{x}-\hat{\mathbf{x}}).$$

Evidently, all trajectories which start on the hyperplane remain there, and all others diverge from it; further, since we are assuming that every eigenvalue of $\mathbf{A}$, apart from λ, has negative real part, the dimension of the stable manifold is one lower than that of the state-space, and hence equal to that of hyperplane, which contains it and which it therefore fills completely. Similarly, the unstable manifold (of the linear model) consists of the line through $\hat{\mathbf{x}}$ in the direction of $\mathbf{w}$, the eigenvector of $\mathbf{A}$ corresponding to λ, since

$$\mathbf{A}\mathbf{w} = \lambda\mathbf{w}$$

which implies that, if $(\mathbf{x}-\hat{\mathbf{x}})$ is proportional to $\mathbf{w}$, then so is $\dot{\mathbf{x}}$. The fact that trajectories diverge from the stable manifold, on both sides, suggests that it constitutes a boundary, or 'separatrix', between regions associated with different kinds of dynamical behaviour; for instance, if there is an asymptotically stable equilibrium point nearby, the separatrix may coincide with the boundary of a domain of attraction, or part of it. Of course, in the actual nonlinear system under consideration, the separatrix will no longer be simply a hyperplane through a type 1 equilibrium point, but may nevertheless be well approximated by it in that vicinity, as shown in Fig. 5.6. Moreover, this method of approximation can be extended, by assuming an expansion in $(\mathbf{x}-\hat{\mathbf{x}})$ for the shape of the true stable manifold, and thus calculating a hypersurface which approximates it.

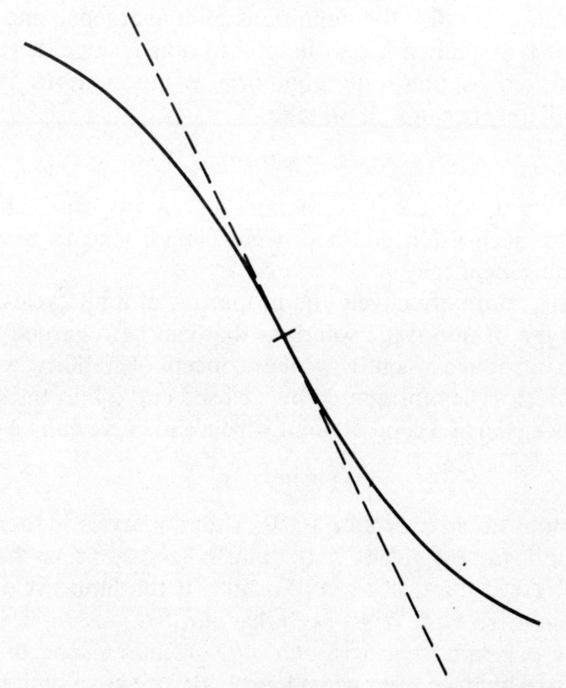

Fig. 5.6 Tangent hyperplane to separatrix through type 1 equilibrium point.

5.5 Stability of motion

So far, we have discussed stability only in relation to equilibrium solutions, but the concept can readily be extended to trajectories in general. Indeed, given any particular solution vector $\mathbf{x}_0(t)$ of an autonomous system, so that

$$\dot{\mathbf{x}}_0 = \mathbf{f}(\mathbf{x}_0)$$

we can define $\tilde{\mathbf{x}} = \mathbf{x} - \mathbf{x}_0$ and rewrite the state-space equations as

$$\dot{\tilde{\mathbf{x}}} = \tilde{\mathbf{f}}(\tilde{\mathbf{x}},t)$$

where

$$\tilde{\mathbf{f}}(\mathbf{x},t) \equiv \mathbf{f}(\tilde{\mathbf{x}}+\mathbf{x}_0(t)) - \mathbf{f}(\mathbf{x}_0(t))$$

with the result that the original motion $\mathbf{x} = \mathbf{x}_0(t)$ now corresponds to the equilibrium point $\tilde{\mathbf{x}} = 0$ of a non-autonomous system. We then define the trajectory $\mathbf{x}_0(t)$ to be *stable* if, given any time t_0 and any $\epsilon > 0$, there exists $\delta > 0$ such that $|\tilde{\mathbf{x}}(t_0)|<\delta$ implies $|\tilde{\mathbf{x}}(t)|<\epsilon$ for all $t>t_0$. This is the natural generalisation of the corresponding definition for a static equilibrium, but is slightly more complicated, in that δ may now depend on t_0, as a consequence of the dynamical equations satisfied by $\tilde{\mathbf{x}}$ being explicitly time-dependent. If it is possible to choose δ, in the above definition, independent of t_0, then the trajectory is said to be *uniformly stable*. In a similar manner, we can generalise the definitions of convergence and asymptotic stability, but it is actually not very helpful to do so, since these properties are never satisfied by non-static equilibria of autonomous systems; this follows immediately because, if we take

$$\mathbf{x}(t) = \mathbf{x}_0(t+\tau)$$

for some fixed $\tau \neq 0$, where $\mathbf{x}_0(t)$ is a periodic or almost-periodic function of t, then $\tilde{\mathbf{x}}(t)$ is also such a function and hence cannot tend to zero as $t \to \infty$ without vanishing identically.

To classify more effectively the properties of limit cycles, which are the simplest type of non-static solutions that can be regarded as dynamic equilibria, we introduce a slightly weaker concept of stability, as follows. A limit cycle trajectory is represented by a closed curve Γ in the state-space, and the distance $\rho(\mathbf{x})$ of a general point $\mathbf{x}$ from this curve can be defined as

$$\rho(\mathbf{x}) = \inf_{\mathbf{q}} |\mathbf{x}-\mathbf{q}|$$

with the infimum taken over all $\mathbf{q}$ on Γ. The limit cycle is then said to be *orbitally stable* if, for any given $\epsilon > 0$, there exists $\delta > 0$ such that $\rho(\mathbf{x}(0)) < \delta$ implies $\rho(\mathbf{x}((t)) < \epsilon$ for all $t > 0$. Further, if the limit cycle is orbitally stable, and also there exists $\Delta > 0$ such that $\rho(\mathbf{x}((0)) < \Delta$ implies $\rho(\mathbf{x}(t)) \to 0$ as $t \to \infty$, then it is said to be *asymptotically orbitally stable*; this is, in fact, the property to which we refer when loosely describing a limit cycle simply as 'stable'. If a limit cycle is not orbitally stable, we can define its stable and unstable manifolds in the same way as for a static equilibrium, and their properties are of some importance in the more subtle ramifications of dynamical system theory; in particular, if they intersect, a very complicated structure arises, which is known to be associated with the appearance of chaos.

Apart from limit cycles, there are also other kinds of trajectory which may be considered as representing a form of nonstatic equilibrium. Since every bounded trajectory possesses a non-empty positive limit set, which is also an invariant set, we can usefully define a 'central' trajectory as one which is contained within its own positive limit set. The simplest central trajectories are equilibrium points and limit cycles, since, in each of these cases, the trajectory actually coincides with the limit set; among more complicated types are almost-periodic trajectories, which again have a quasi-equilibrium character, even though they are not closed curves. In fact, any central trajectory which is stable can be shown to possess a kind of recurrence property, in the following sense. If $\mathbf{x}(t)$ denotes such a trajectory, then, because each point $\mathbf{x}(t_0)$ on it belongs to its positive limit set, there exists, for any $\delta > 0$, a sequence $t_k \to \infty$ such that

$$|\mathbf{x}(t_0) - \mathbf{x}(t_k)| < \delta$$

for every positive integer k; consequently, since the trajectory is stable, we can choose δ, given any $\epsilon > 0$, so that

$$|\mathbf{x}(t_0+\tau) - \mathbf{x}(t_k+\tau)| < \epsilon$$

for all $\tau > 0$. This is rather a weak form of recurrent behaviour, encompassing more than just the periodic and almost-periodic types; its main significance, however, is that a central trajectory not possessing it must be unstable, which appears to characterise chaotic motion.

As regards the local stability properties of general trajectories, the natural approach to follow is linearisation around the nominal solution. For a trajectory $\mathbf{x}_0(t)$, this yields

$$\dot{\tilde{\mathbf{x}}} = \mathbf{A}(t)\tilde{\mathbf{x}}$$

where the coefficient matrix $\mathbf{A}(t)$ is given by

$$\mathbf{A}(t) = \nabla_{\mathbf{x}}\mathbf{f}(\mathbf{x}_0(t))$$

which is, in general, time-dependent. If the trajectory is a limit cycle, of period T, then $\mathbf{A}(t)$ will clearly also be periodic, with

$$\mathbf{A}(t+T) = \mathbf{A}(t)$$

for all t; similarly, if the trajectory is almost-periodic, then so is $\mathbf{A}(t)$. In the periodic case, stability can be assessed by using the 'discrete-time' relation

$$\tilde{\mathbf{x}}((k+1)T) = \Phi(T,0)\,\tilde{\mathbf{x}}(kT)$$

which follows since the transition matrix $\Phi(t,t_0)$ satisfies the same equations as $\Phi(t+T,t_0+T)$, on account of the periodicity, and k is an integer. The linearised system is thus stable if all eigenvalues of $\Phi(T,0)$ lie in the closed unit circle (and any on the boundary are distinct); in fact, one of them is always unity (corresponding to motion along the trajectory), and the system is asymptotically orbitally stable if the rest are in the open unit circle.

Example 5.3 Stability and orbital stability

Consider a system described by the rather artificial set of state-space equations

$$\dot{x}_1 = x_2\sqrt{(x_1^2 + x_2^2)}$$
$$\dot{x}_2 = -x_1\sqrt{(x_1^2 + x_2^2)}$$

whose general solution is given by

$$x_1 = r\sin(rt + \phi)$$
$$x_2 = r\cos(rt + \phi)$$

where r and ϕ are constants, determined from the initial conditions, so that the trajectories are circles, centred on the origin, as in Fig. 5.7. It is clear that every trajectory is orbitally stable though not asymptotically so, since a point moving on any given trajectory remains at a fixed distance from any other. On the other hand, no trajectory (apart from the degenerate case $\mathbf{x} \equiv 0$), is stable, because circles of different radii are traversed at different angular velocities; hence, the distance between representative points (on trajectories with $r = r_1$ and $r = r_2$) varies periodically between $(r_1 + r_2)$ and $|r_1 - r_2|$, so that it cannot be made to remain arbitrarily small by choosing the initial positions sufficiently close together.

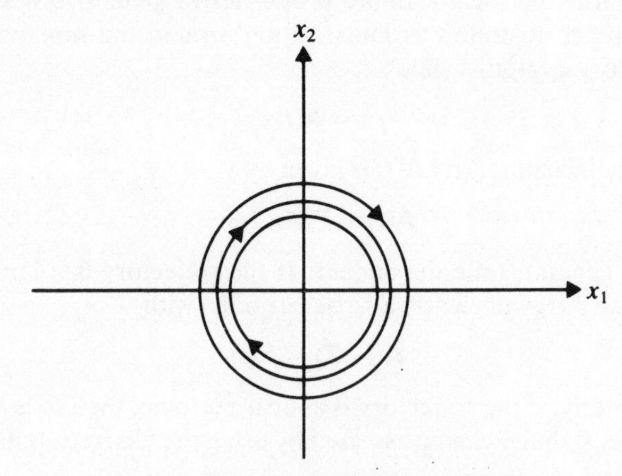

Fig. 5.7 Trajectories for Example 5.3.

5.6 Time-varying systems

For a system described by a set of time-dependent state-space equations

$$\dot{\mathbf{x}} = \mathbf{f}(\mathbf{x},t)$$

Lyapunov's second method can still be applied, with certain modifications. The Lyapunov function may now be an explicit function of time $V(\mathbf{x},t)$, so that

$$\dot{V}(\mathbf{x},t) = \frac{\partial V}{\partial t} + \sum_{j=1}^{n} \frac{\partial V}{\partial x_j} f_j(\mathbf{x},t)$$

where the partial derivatives are assumed to be continuous in t as well as $\mathbf{x}$. Since we can, in principle, transform any solution into an equilibrium at the origin, by a change of variables, we will suppose this to have been done, that is to say, we take

$$\mathbf{f}(0,t) = 0$$

for all t, and correspondingly

$$V(0,t) \equiv 0.$$

The derivation of stability conditions then follows the same lines as in the autonomous case, with a slight strengthening of assumptions. Specifically, we require the existence of a continuous scalar function $V_0(\mathbf{x})$, with $V_0(0) = 0$, such that, for all t,

$$V(\mathbf{x},t) \geqslant V_0(\mathbf{x})$$

whenever $\mathbf{x}$ is in some open connected set S which contains the origin, where $V_0(\mathbf{x})$ satisfies the conditions of positivity, with respect to S, and (if necessary) radial unboundedness, as previously defined. Then, if we also have

$$\dot{V}(\mathbf{x},t) \leqslant 0$$

for all $\mathbf{x}$ in S and all t, it follows, by a straightforward extension of the argument in the time-invariant case, that the equilibrium is stable; moreover, if S is the entire state-space and V_0 is radially unbounded, then all trajectories remain bounded. In order to show that these properties hold uniformly with respect to time, however, we need a further condition on the Lyapunov function, known as decrescency: $V(\mathbf{x},t)$ is said to be decrescent in S if there exists a continuous positive function $U(\mathbf{x})$ with $U(0) = 0$, such that

$$V(\mathbf{x},t) \leqslant U(\mathbf{x})$$

for all $\mathbf{x}$ in S and all t. Also, to establish asymptotic stability, we further require that $\dot{V}$ satisfies the condition

$$\dot{V}(\mathbf{x},t) \leqslant W(\mathbf{x})$$

for all t, where $W(\mathbf{x})$ is some continuous function which is negative for all $\mathbf{x}(\neq 0)$ in S.

Example 5.4 A bilinear system

Suppose the state-space equations are

$$\dot{x}_1 = x_2$$
$$\dot{x}_2 = -u_1 x_1 - u_2 x_2$$

where the input components (u_1,u_2) are specified as functions of time. By analogy with the time-invariant case, where u_1 and u_2 would be spring and damping constants, respectively, we take the Lyapunov function

$$V = x_1^2 + \frac{x_2^2}{u_1}$$

whence

$$\dot{V} = -\left(\frac{2u_2}{u_1} + \frac{\dot{u}_1}{u_1^2}\right) x_2^2$$

from the state-space equations. Provided that u_1 is always positive and has a finite upper bound, and that

$$\dot{u}_1 + 2u_1 u_2 \geqslant 0$$

for all t, it is clear that V and $\dot{V}$ satisfy the conditions required for us to conclude that the system is stable and all trajectories are bounded. Further, if u_1 also has a positive lower bound, then V is decrescent and so the system is uniformly stable, though not necessarily asymptotically stable.

5.7 Feedback system stability

The stability problems encountered in control engineering are usually associated with the behaviour of feedback systems, and considerable attention has therefore been devoted to this specific area. One can, indeed, study any dynamical system in this context; a convenient way of doing so, formally at least, is to separate the part which can be described by linear time-invariant equations from everything else, so that the system is represented as a feedback connection of two blocks, as in Fig. 5.8. Introducing a state-space representation for the linear time–invariant part,

$$\dot{\mathbf{x}} = \mathbf{A}\mathbf{x} + \mathbf{B}\mathbf{u}$$
$$\mathbf{y} = \mathbf{C}\mathbf{x}$$

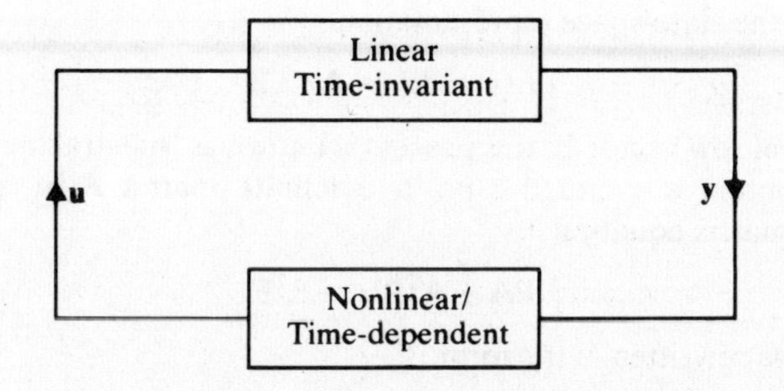

Fig. 5.8 Feedback connection of two subsystems.

we then close the feedback loop by defining the relation, which may be nonlinear and/or time-dependent, between **u** and **y**. For simplicity, we will here consider only the case of a single-input, single-output relation, that is to say, we take **u** and **y** as scalar variables, and suppose that

$$u(t) = -F(\mathbf{x},t)\, y(t)$$

so that $F(\mathbf{x},t)$ represents a variable feedback gain, the minus sign being introduced for consistency with the usual negative feedback convention. In view of the widespread use of frequency-response methods in control system design, it is natural to look for stability conditions which can be expressed in terms of the transfer function

$$G(s) = \mathbf{C}(s\mathbf{I}-\mathbf{A})^{-1}\,\mathbf{B}$$

for the linear time-invariant block. With this in mind, we introduce the concept of a 'positive-real' transfer function: $G(s)$ is said to be positive-real if it has no poles in the open right half-plane, any poles on the imaginary axis are simple, and the real part is non-negative at all points, apart from poles, on the imaginary axis. This is equivalent to the statement that the linear system with transfer function $G(s)$ is stable and

$$G(\mathrm{i}\omega) + G(-\mathrm{i}\omega) \geqslant 0$$

for all real ω. If this condition holds, and if we also have

$$F(\mathbf{x},t) \geqslant 0$$

for all $\mathbf{x}$ and t, then stability of the feedback system can be proved, by either Lyapunov or input–output methods.

A rigorous proof of the existence of a suitable Lyapunov function depends on certain concepts of linear (algebraic) system theory, lying outside the scope of this book, but an outline of the procedure is as follows. Because $G(s)$ is positive-real, we can construct a rational function $L(s)$ satisfying

$$L(s)L(-s) = G(s) + G(-s)$$

and having the state-space representation

$$L(s) = \mathbf{E}(s\mathbf{I}-\mathbf{A})^{-1}\,\mathbf{B}$$

for some real row vector $\mathbf{E}$; this process is known as 'spectral factorisation'. We next obtain a symmetric positive-definite matrix $\mathbf{P}$ by solving the Lyapunov matrix equation

$$\mathbf{PA} + \mathbf{A}^{\mathrm{T}}\mathbf{P} = -\mathbf{E}^{\mathrm{T}}\mathbf{E}$$

which can be rewritten in the form

$$\mathbf{P}(s\mathbf{I}-\mathbf{A}) + (-s\mathbf{I}-\mathbf{A}^{\mathrm{T}})\mathbf{P} = \mathbf{E}^{\mathrm{T}}\mathbf{E}$$

whence, postmultiplying by $(s\mathbf{I}-\mathbf{A})^{-1}\,\mathbf{B}$ and premultiplying by $\mathbf{B}^{\mathrm{T}}(-s\mathbf{I}-\mathbf{A}^{\mathrm{T}})^{-1}$, we obtain

$$\mathbf{B}^{\mathrm{T}}\mathbf{P}(s\mathbf{I}-\mathbf{A})^{-1}\mathbf{B} + \mathbf{B}^{\mathrm{T}}(-s\mathbf{I}-\mathbf{A}^{\mathrm{T}})^{-1}\mathbf{PB} = L(-s)L(s)$$

so that we can identify

$$\mathbf{B}^{\mathrm{T}}\mathbf{P} = \mathbf{C}$$

from the expression for $G(s)$. Then, using the Lyapunov function

$$V(\mathbf{x}) = \mathbf{x}^{\mathrm{T}}\mathbf{P}\mathbf{x}$$

we obtain from the state-space equations

$$\begin{aligned}\dot{V}(\mathbf{x},t) &= -\mathbf{x}^{\mathrm{T}}\mathbf{E}^{\mathrm{T}}\mathbf{E}\mathbf{x} - 2\mathbf{x}^{\mathrm{T}}\mathbf{C}^{\mathrm{T}}F(\mathbf{x},t)\mathbf{C}\mathbf{x} \\ &= -(\mathbf{E}\mathbf{x})^2 - 2\,(\mathbf{C}\mathbf{x})^2\,F(\mathbf{x},t) \leqslant 0\end{aligned}$$

so that the equilibrium at $\mathbf{x} = 0$ is stable and all the trajectories of the feedback system are bounded; moreover, since V is decrescent, these properties hold uniformly in t.

The assumptions made above are not sufficient to establish asymptotic stability, but this can also be achieved if we strengthen the condition on $G(s)$ to 'strict positive-realness', which means that all its poles are in the open left half-plane, and the real part is positive everywhere on the imaginary axis. Also, it is not necessary to assume, as we have implicitly done, that $G(s) \to 0$ as $s \to \infty$; the same results still hold if the limiting value is finite and non-zero, though the analysis then becomes somewhat more complicated. Further, if $G(s)$ is positive-real, it is always possible to find an arbitrarily small constant $\delta > 0$ such that the rational function $\{G(s) + \delta\}/\{1 + \delta G(s)\}$ is strictly positive-real.

In practice, it is not very likely that either $G(s)$ or $F(\mathbf{x},t)$ will satisfy the conditions which we have invoked. However, the results can be made more readily applicable by exploiting the freedom to make certain changes of dynamical variables. Thus, if $F(\mathbf{x},t)$ instead satisfies the restriction

$$\alpha \leqslant F(\mathbf{x},t) < \beta$$

for all **x** and t, where α and β are some constants, it is convenient to define new variables $\tilde{u}$ and $\tilde{y}$ by setting

$$u = \beta\tilde{u} - \alpha\tilde{y}$$
$$y = \tilde{y} - \tilde{u}$$

so that

$$\tilde{u}(t) = -\tilde{F}(\mathbf{x},t)\,\tilde{y}(t)$$

with

$$\tilde{F}(\mathbf{x},t) = \frac{F(\mathbf{x},t)-\alpha}{\beta-F(\mathbf{x},t)} \geqslant 0$$

and the transfer function of the linear block with input $\tilde{u}$ and output $\tilde{y}$ is

$$\tilde{G}(s) = \frac{1+\beta G(s)}{1+\alpha G(s)}$$

where the relationship between the original and transformed systems is illustrated in Fig. 5.9. Hence, if $\tilde{G}(s)$ is positive-real, it follows that the

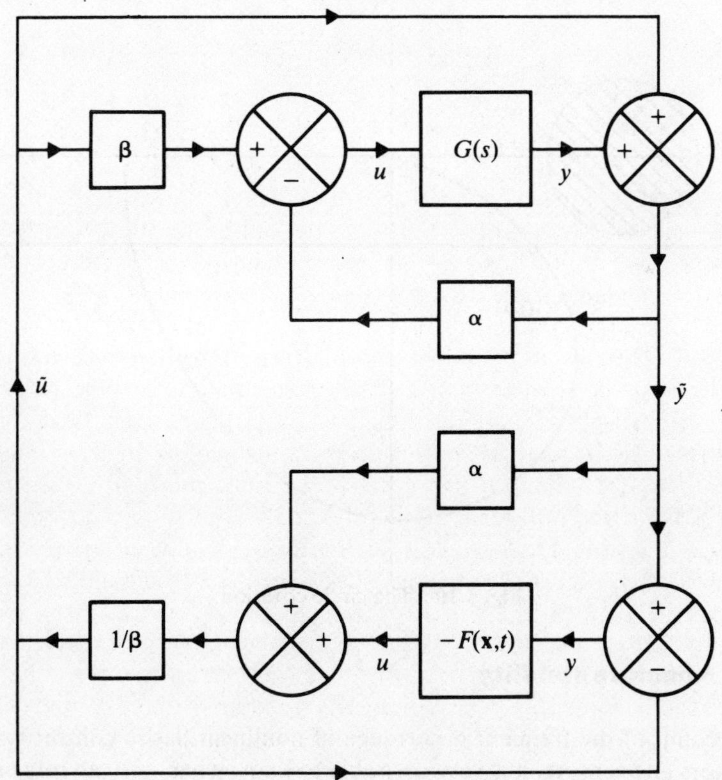

Fig. 5.9 System transformation.

equilibrium of the feedback system at $\mathbf{x} = 0$ is uniformly stable, as before; moreover, if $\tilde{G}(s)$ is strictly positive-real, the system is globally asymptotically stable. Results of this kind, which show that a system remains stable so long as a varying parameter stays within specified bounds, are sometimes called 'absolute stability criteria'; the term 'hyperstability' is also used. A useful feature of this type of criterion is the existence of a graphical interpretation, in terms of Nyquist diagrams. Assuming that $G(s)$ is the transfer function of an asymptotically stable system, the necessary and sufficient condition for $\tilde{G}(s)$ to be strictly positive-real is that the plot of $G(i\omega)$, for all real ω, has one of the following properties: if $\alpha\beta < 0$, it lies inside the disc on $(-1/\beta, -1/\alpha)$ as diameter; if $\alpha = 0$ or $\beta = 0$, it lies entirely to the right of $-1/\beta$ or the left of $-1/\alpha$, respectively; if $\alpha\beta > 0$, it lies outside, and does not encircle, the disc on $(-1/\alpha, -1/\beta)$ as diameter. The case with $\beta > \alpha > 0$ is illustrated in Fig. 5.10; it is often referred to as the 'circle criterion', since the graph of $G(i\omega)$ is required to avoid, and not encircle, a 'critical disc' which generalises the classical Nyquist critical point.

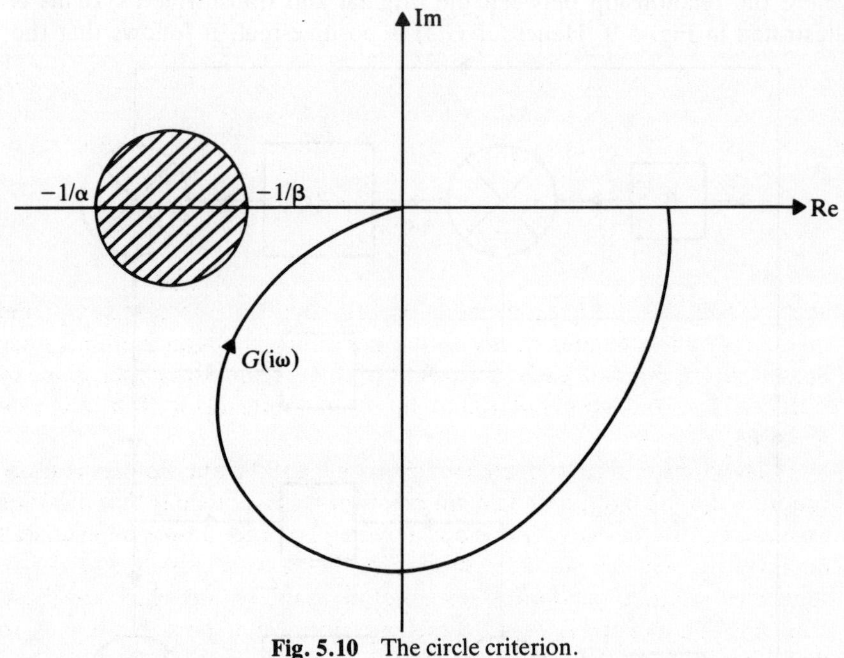

Fig. 5.10 The circle criterion.

5.8 Absolute stability

On account of the frequent occurrence of nonlinearities in control system actuators and sensors, it is very desirable to know when one can rely on the system being stable, even though it has been designed on the tacit assumption that these devices are linear. The simplest case, involving only one nonlinear element, may be represented as in Fig. 5.8, with a time-invariant feedback relation

$$u = -f(y)$$

so that

$$\dot{\mathbf{x}} = \mathbf{A}\mathbf{x} - \mathbf{B}f(\mathbf{C}\mathbf{x})$$

for some function $f(\cdot)$, satisfying known constraints. Under these circumstances, it was conjectured long ago by Aizerman that this nonlinear system would be stable if the linear system

$$\dot{\mathbf{x}} = (\mathbf{A} - \mathbf{B}\lambda\mathbf{C})\mathbf{x}$$

were asymptotically stable for every constant λ in the range over which the 'effective gain' $f(y)/y$ is permitted to vary; also, a similar conjecture was made by Kalman, with the derivative $\partial f(y)/\partial y$ replacing the effective gain. Unfortunately, however, though they are often valid in practice, it has been shown by means of counterexamples that both these conjectures are, in general, false. Similarly, the even more naive idea that a describing function analysis might be sufficient to prove stability is also untrue, not surprisingly in view of its approximate nature. We do, nevertheless, have the circle criterion, which could be applied in this case, with the nonlinearity satisfying (for $y \neq 0$)

$$\alpha < \frac{f(y)}{y} < \beta$$

and the Nyquist plot lying in the appropriate part of the complex plane; this, incidentally, relates neatly to the describing function method, since the describing function can be shown to lie in the region which the graph of $-1/G(i\omega)$ is forbidden to enter, and thus cannot intersect it, if the criterion is satisfied.

The circle criterion, of course, gives only a sufficient, not a necessary, condition for stability, and its main practical disadvantage is that it can be extremely conservative. It turns out, however, that for a time-independent nonlinearity such as we are considering here, it is possible to derive somewhat sharper conditions for absolute stability, of which the best-known is due to Popov. To state this criterion, we assume that $\alpha = 0$, so that the constraint

$$0 < yf(y) < \beta y^2$$

is satisfied for $y \neq 0$. Popov's result is then that, if there exists a constant γ such that the rational function

$$\tilde{G}(s) = (1+\gamma s)\, G(s) + \frac{1}{\beta}$$

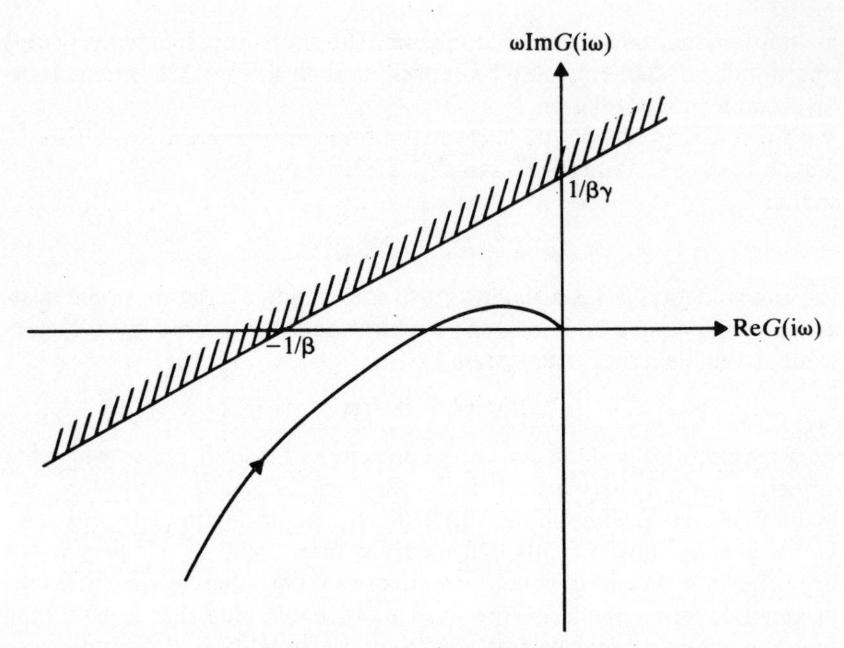

Fig. 5.11 The Popov criterion.

is positive-real, then the feedback system is globally asymptotically stable. This can be proved, for $\gamma \geqslant 0$, by using a Lyapunov function of the form

$$V = \mathbf{x}^{\mathrm{T}}\mathbf{P}\mathbf{x} + 2\gamma \int_0^y f(v)\,\mathrm{d}v$$

where $\mathbf{P}$ is obtained from spectral factorisation of $\{\tilde{G}(s) + \tilde{G}(-s)\}$, and the result can then be extended to $\gamma < 0$ through a change of variables, although this requires $f(y)$ to be single-valued. Like the circle criterion, this result has a graphical interpretation, since

$$\mathrm{Re}\,\tilde{G}(\mathrm{i}\omega) = \mathrm{Re}\,G(\mathrm{i}\omega) - \gamma\omega\,\mathrm{Im}\,G(\mathrm{i}\omega) + \frac{1}{\beta}$$

which enables the positive-realness condition to be checked by means of the 'modified polar plot' of $\omega\mathrm{Im}G(\mathrm{i}\omega)$ against $\mathrm{Re}G(\mathrm{i}\omega)$, as in Fig. 5.11; the graph is required to lie on the right of some straight line (of slope $1/\gamma$) passing through the point $-1/\beta$.

Example 5.5 Comparison of stability conditions

In order to compare the circle and Popov criteria with other indicators of stability properties, let us re-examine Example 4.1, using a smoothed version of the discontinuous nonlinear function: we can take, for instance,

$$f(y) = \frac{h}{2}\left\{\tanh\left(\frac{y+b}{\epsilon}\right) + \tanh\left(\frac{y-b}{\epsilon}\right)\right\}$$

with a constant $\epsilon > 0$, and recover the three-level relay characteristic of Fig. 4.3 as $\epsilon \to 0$. With $\epsilon > 0$, as in Fig. 5.12, we have

$$0 < \frac{f(y)}{y} < \frac{h}{b}$$

for $y \neq 0$, and the absolute stability criteria can be applied. The Nyquist plot, obtained from

$$G(\mathrm{i}\omega) = \frac{1 - \mathrm{i}K\omega}{\mathrm{i}\omega(1+\mathrm{i}\omega)}$$

crosses the real axis at $-K$, as illustrated in Fig. 5.13, and lies entirely to the right of the vertical line through $-(K+1)$, which is its asymptote as $\omega \to 0$. Consequently, the circle criterion, corresponding to Popov with $\gamma = 0$, gives global asymptotic stability for $K + 1 \leqslant b/h$. However, taking $\gamma = 1$ instead, we get

$$\tilde{G}(s) = \frac{1}{s} - K + \frac{b}{h}$$

and so the Popov criterion gives the stronger result that the system is globally asymptotically stable for $K \leqslant b/h$. Evidently, Aizerman's conjecture is correct in this case, since the system would be asymptotically stable with the nonlinear element replaced by any constant gain strictly between 0 and h/b, if $Kh \leqslant b$.

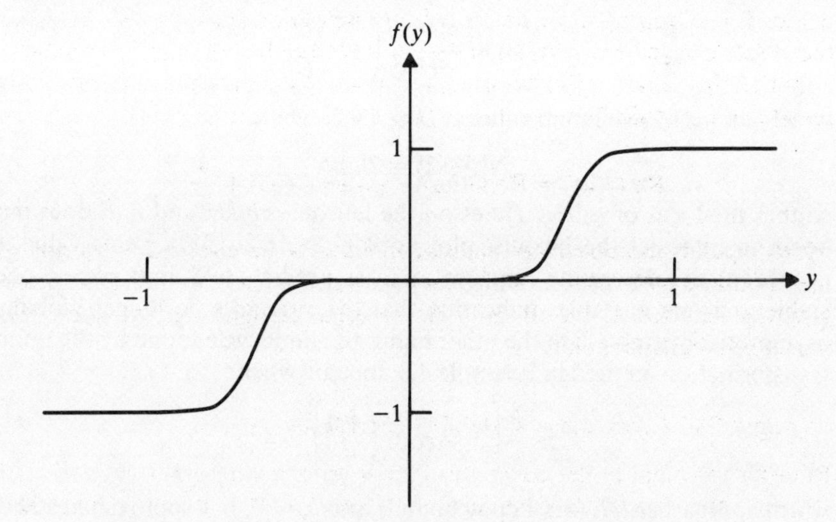

Fig. 5.12 Nonlinearity in Example 5.5 ($h = 1, b = 0.6, \epsilon = 0.1$).

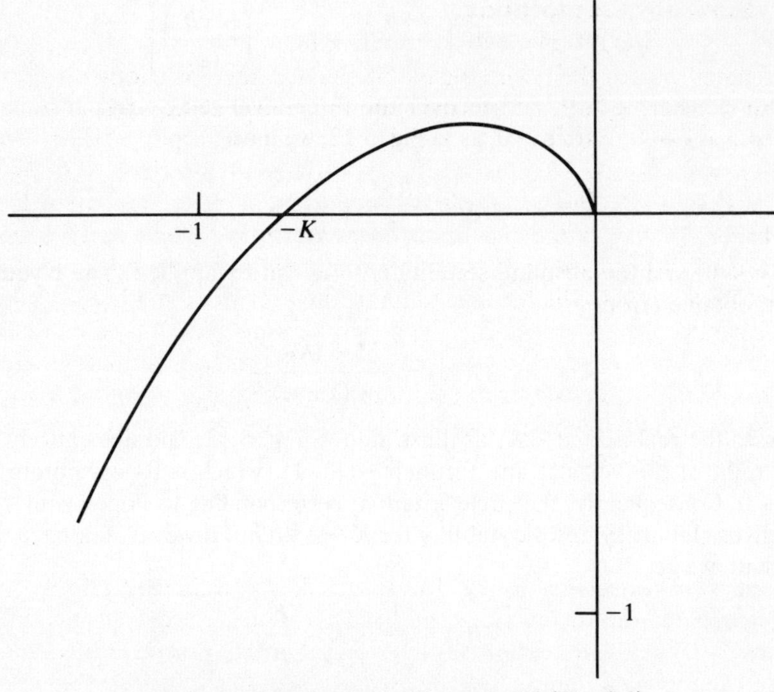

Fig. 5.13 Nyquist plot for Example 5.5 ($K = 0.8$).

For an approximate investigation of how the stability breaks down as K is further increased, we turn to the describing function method. In the limit $\epsilon \to 0$, the SIDF is given for $U \geqslant b$ by

$$N(U) = \frac{4h}{\pi U}\sqrt{\left(1 - \frac{b^2}{U^2}\right)}$$

which attains its maximum value at $U = b\surd 2$, where

$$N(b\surd 2) = 2h/\pi b$$

so that the locus of $-1/N(U)$ lies on the left of $-\pi b/2h$, and thus does not begin to intersect the Nyquist plot until $K \geqslant \pi b/2h$. When this value is exceeded, the describing function analysis predicts two limit cycles, one stable and one unstable, indicating that the system is no longer globally asymptotically stable. On the other hand, the limit cycle found by the point transformation method in Example 4.1 appears when

$$K > \frac{b}{h}\left\{1 + \coth\left(\frac{b}{h}\right)\right\} - 1$$

which approaches b/h (the Popov bound) as $b/h \to 0$, but approximates $2b/h$ as $b/h \to \infty$.

5.9 Input–output methods

As was mentioned at the beginning of this chapter, there is another way of defining stability, different from the state-space formulation but in some respects complementary to it, by means of the input–output relation. To use this in the context of feedback systems, we need to introduce, as in Fig. 5.14, an external input r, which can represent a reference signal, a disturbance, or simply the zero-input response of the linear part of the system, since it may not have been in its null state ($\mathbf{x} = 0$) at the initial instant. Each block in the feedback loop is then regarded mathematically as an operator which transforms input signals into output signals, and stability is defined in terms of the properties of such transformations. In order to do this, we have to introduce norms on the signals; there are many possibilities, but the one most closely related to the Lyapunov approach is the so-called L_2-norm

$$||u|| = \sqrt{\left(\int_0^\infty |u(t)|^2 \, \mathrm{d}t\right)}$$

for a scalar or vector signal $u(t)$. The feedback system is then said to be finite-gain input–output stable if the norms of its internal signals are bounded by some fixed multiple of the external input norm. If this is true, then a Lyapunov function can be constructed, to estabish global asymptotic stability.

A further advantage of the L_2-norm is that it can be used to characterise the properties of a linear system in terms of its frequency response, through Fourier transforms and Parseval's theorem. Thus, for a stable single-input, single-output system with transfer function $G(s)$, the supremum of the ratio of output norm to input norm is

$$\sup_u \frac{||y||}{||u||} = \sup_\omega |G(\mathrm{i}\omega)|$$

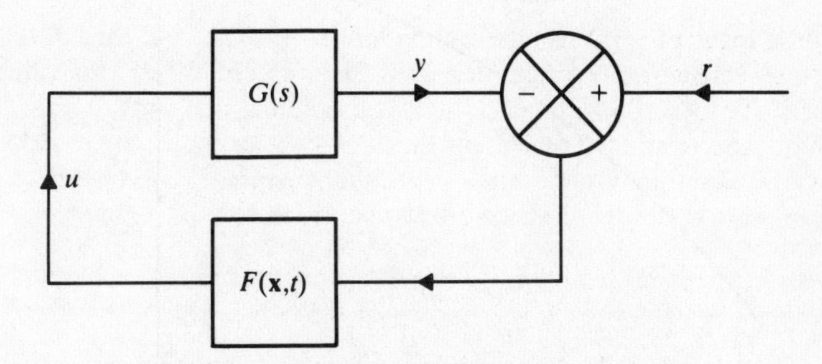

Fig. 5.14 Feedback loop with external input.

known in this context as the system gain. Also, such a system is said to be 'passive' if we always have

$$\int_0^\infty u(t)\, y(t)\, dt \geqslant 0$$

and ths is equivalent to $G(s)$ being positive-real, as well as to the relationship

$$||u-y|| \leqslant ||u+y||$$

between the norms. By using these concepts, one can obtain absolute stability results, including the circle and Popov criteria, in an input–output setting. In the case of the Popov criterion, it is necessary to introduce certain fictitious operators, known as 'multipliers', into the feedback loop and, by an extension of this technique, it is possible to obtain many other stability criteria, more readily than by constructing Lyapunov functions.

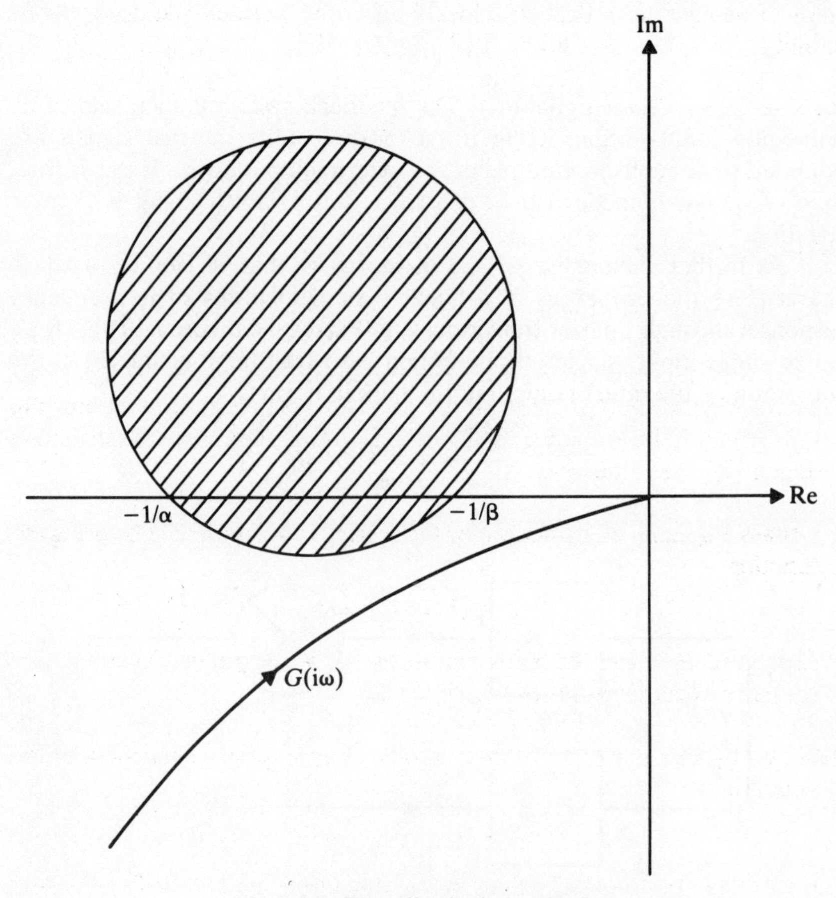

Fig. 5.15 The off-axis circle criterion.

One such result, applicable for a single-valued time-invariant nonlinearity satisfying

$$\alpha < \frac{\partial f(y)}{\partial y} < \beta$$

is the 'off-axis' circle criterion, where the plot of $G(i\omega)$, for $\omega \geqslant 0$, is required to avoid an arbitrary circle passing through $-1/\alpha$ and $-1/\beta$, as illustrated in Fig. 5.15. This result bears the same kind of relationship to Kalman's conjecture as Popov's criterion does to the Aizerman conjecture.

The input-output approach is also in some ways more convenient than state-space methods for application to multi-input, multi-output systems, as well as to more complicated interconnection structures than the simple two-element loop considered here. General interconnections, however, can alternatively be handled by combining Lyapunov functions for the subsystems, and the Lyapunov method in any case has the further merit of being able to deal with local properties, without requiring global stability.

5.10 Exercises

1 For the first-order system

$$\dot{x} = x - x^3 + r$$

where x is a scalar variable and r is an adjustable constant, find how the stable equilibrium points and their maximal domains of attraction depend on the value of r.

2 Obtain a domain of attraction for the system in Example 5.2 by using the function

$$V = y^2 - \tfrac{2}{3} y^3 + \dot{y}^2$$

expressed in terms of state variables, as a Lyapunov function, and compare with the results of Example 5.2.

3 Show, by Lyapunov or other methods, that a system described by the equation

$$\ddot{y} + h(\dot{y}) + f(y) = 0$$

is globally asymptotically stable if the functions f and h satisfy $yf(y) \geqslant \epsilon y^2$ for some $\epsilon > 0$ and $\dot{y}\, h(\dot{y}) > 0$ for all $\dot{y} \neq 0$.

4 Find what information can be obtained about the stability of the system in Example 5.4 by using

$$V = u_1 x_1^2 + x_2^2$$

as an alternative Lyapunov function.

5 Draw the modified polar plot for the transfer function $1/s(s+1)$ and hence apply the Popov criterion to a feedback system whose linear element has this transfer function.

6 CONTROL SYSTEM PERFORMANCE

To the extent that a plant can be described by a linear time-invariant model, there are many well-established techniques which can be used to design a control system for it. Even if the plant is known to be nonlinear, the initial design is usually based on a linearised model, and one then has the problem of assessing how the actual nonlinearities will affect its performance. Also, especially if the deterioration is significant, it is desirable to know any modifications which might be introduced in order to counteract the nonlinear effects. In this chapter, we shall try to cover both these aspects.

So far as the assessment of performance is concerned, the only way to determine the behaviour of the system in detail, short of practical implementation, is by simulation on a computer. On the other hand, this is not necessarily the best way to proceed initially, since a particular simulation run only gives information corresponding to the conditions of operation assumed, and so may need to be repeated many times, under different conditions, in order to give an adequate picture of the performance of a nonlinear system. Consequently, it is also useful to have alternative approaches which can provide information about the sytem's behaviour over a range of possible conditions; in this regard, we shall consider rigorous methods for putting bounds on the values of dynamical variables, as well as techniques of approximate analysis based on the describing function idea.

If the nonlinear nature of the system turns out to cause a serious degradation in performance, there are a number of possible ways to deal with it, which we shall discuss. One approach is to inject a rapidly oscillating signal, known as 'dither', at the input to a nonlinear element, with the aim of effectively smoothing it out, so that the system becomes more nearly linear in its overall behaviour. Another possibility is simply to implement a nonlinear control law, chosen so as to compensate directly for the nonlinearity of the plant, but this is fully practicable only in certain rather special cases. It may be remarked, incidentally, that the application of feedback control normally tends to diminish, rather than enhance, any nonlinear effects which happen to be present, provided that the system remains adequately stable; however, the maintenance of stability can itself be one of the difficulties caused by nonlinearities. This problem, in fact, typically manifests itself when the system is required to operate across a wide range of external circumstances, giving the nonlinear features more opportunity to show up. A common procedure in such cases is to use a set of different linear controllers, based on linearised models corresponding to various operating points, and bring them into action according to a prescribed schedule as the ambient conditions and demand signals change. Actually, this is a rudimentary form of what is called adaptive control, although the term is more generally used to describe the type of scheme in which the control parameters are altered in accordance with the observed behaviour of the system, so that it can, in effect, monitor and improve its own performance as it operates.

6.1 Bounds on system variables

The following discussion will be based on the simple nonlinear feedback control loop shown in Fig. 6.1, where the objective is to make the output $y(t)$, of the linear system with transfer function $G(s)$, follow the reference signal $r(t)$ as closely as possible, subject to the effect of the disturbance $q(t)$. With the error signal denoted by

$$e = r - y$$

the input to the linear element becomes

$$u = f(e) + q$$

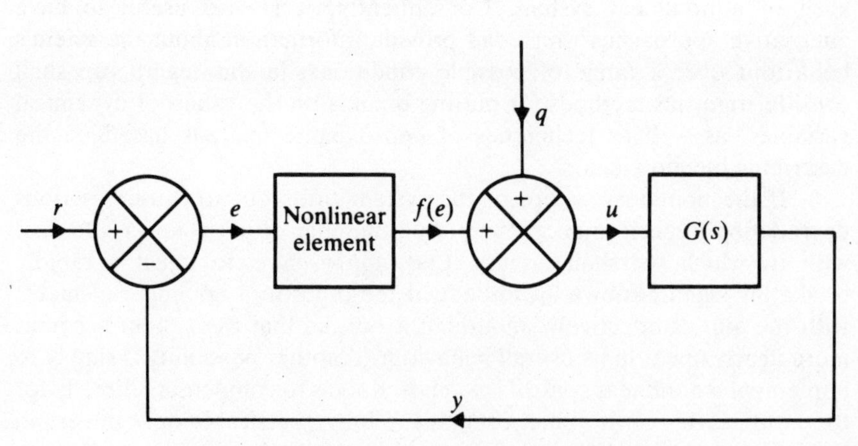

Fig. 6.1 A nonlinear feedback control system.

where the nonlinear function $f(e)$ can represent, in general, either sensor or actuator nonlinearities, or possibly a combination of both. We suppose that it can be approximated by a linear function, so that

$$f(e) = Ke + \Delta f(e)$$

where K is constant and $\Delta f(e)$ satisfies some kind of constraint, to be specified later. Next, we introduce a state-space representation for $G(s)$, giving

$$\begin{aligned} \dot{\mathbf{x}} &= \mathbf{A}\mathbf{x} + \mathbf{B}u \\ y &= \mathbf{C}\mathbf{x} \end{aligned}$$

where, in the interests of simplicity and realism, we have assumed that the plant has no instantaneous term in its input–output relation. Combining the above equations, we get, for the closed-loop system,

$$\begin{aligned} \dot{\mathbf{x}} &= (\mathbf{A}-\mathbf{B}K\mathbf{C})\mathbf{x} + \mathbf{B}(Kr+q) + \mathbf{B}\Delta f(e) \\ e &= r - \mathbf{C}\mathbf{x} \end{aligned}$$

from which we can derive an integral equation for the error. To this end, we define the closed-loop transfer function from u to y, for the linear system obtained by setting $\Delta f \equiv 0$, namely

$$G_K(s) = \frac{G(s)}{1+G(s)K} = \mathbf{C}(s\mathbf{I} - \mathbf{A} + \mathbf{B}K\mathbf{C})^{-1}\mathbf{B}$$

with the corresponding impulse response function

$$g_K(t) = \mathbf{C}\exp\{(\mathbf{A}-\mathbf{B}K\mathbf{C})t\}\mathbf{B}$$

and then, solving the state-space equations, we obtain

$$e(t) = v(t) - \int_0^t g_K(t-\tau)\, \Delta f(e(\tau))\mathrm{d}\tau$$

where $v(t)$ is a driving term arising from the reference input, disturbance and initial condition effects. Specifically, if we let $\tilde{r}$ denote the output of a linear system with input r and transfer function

$$1 - G_K(s)K = \frac{1}{1+G(s)K}$$

so that

$$\tilde{r}(t) = r(t) - \int_0^t g_K(t-\tau)\, Kr(\tau)\, \mathrm{d}\tau$$

and also let

$$\tilde{q}(t) = \int_0^t g_K(t-\tau)\, q(\tau)\, \mathrm{d}\tau$$

and

$$v_0(t) = \mathbf{C}\exp\{(\mathbf{A}-\mathbf{B}K\mathbf{C})t\}\,\mathbf{x}(0)$$

then

$$v(t) = \tilde{r}(t) - \tilde{q}(t) - v_0(t).$$

In the absence of any disturbance and of any deviation from linearity, that is to say, with $q(t) \equiv 0$ and $f(e) \equiv Ke$, it is clear that we should have $e(t) = \hat{e}(t)$, where

$$\hat{e}(t) = \tilde{r}(t) - v_0(t)$$

which is the error signal generated by the linear closed-loop system in response to the given reference input, from the given initial state. Now, let us suppose that the disturbance is bounded, so that, for some constant μ, we have

$$|q(t)| \leqslant \mu$$

for all $t \geqslant 0$, and that $\Delta f(e)$ satisfies the condition that there is a constant ρ such that

$$|\Delta f(e)| \leqslant \rho|e|$$

for all e. This assumption is actually less restrictive than it may appear, since a bounded, but possibly discontinuous or even multivalued, part of the nonlinearity can always be incorporated into q instead of $f(e)$. Then, rewriting the integral equation for $e(t)$ as

$$e(t) = \hat{e}(t) - \int_0^t g_K(t-\tau)\{q(\tau) + \Delta f(e(\tau))\}\mathrm{d}\tau$$

and defining

$$\gamma(t) = \int_0^t |g_K(\tau)|\mathrm{d}\tau$$

we obtain, by taking moduli,

$$|e(t)-\hat{e}(t)| \leqslant \gamma(t)\,\{\mu+\rho \sup_{0<\tau<t} |e(\tau)|\}$$

from which it follows (using the triangle inequality $|e| \leqslant |\hat{e}| + |e-\hat{e}|$) that

$$|e(t)-\hat{e}(t)| \leqslant \frac{\gamma(t)}{1-\rho\gamma(t)}\,\{\mu+\rho \sup_{0<\tau<t} |\hat{e}(\tau)|\}$$

for all t such that

$$\gamma(t) < 1/\rho.$$

The point of this result is that it gives an explicit bound (illustrated in Fig. 6.2) on the deviation of the error from the value it would have in the linear

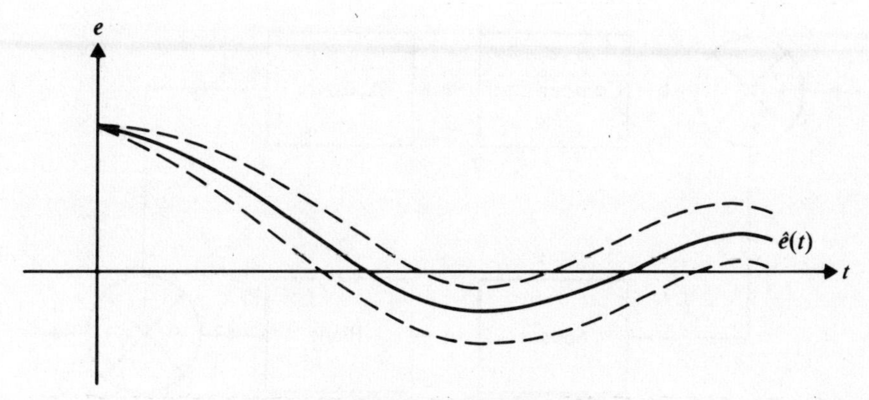

Fig. 6.2 Bounds on the error signal.

case, which can be computed entirely from linear system analysis, given the values of μ and ρ. We note, in particular, that if $\gamma(t) \rightarrow \gamma(\infty)$ as $t \rightarrow \infty$, with

$$\rho\gamma(\infty) < 1$$

then the error signal and all other system variables remain bounded provided that the reference and disturbance inputs do so; the system is then said to possess bounded-input bounded-output stability. Also, if we define a norm (the L_∞-norm) of a signal as the supremum of its modulus over all time, then ρ and $\gamma(\infty)$ can be interpreted as subsystem gains, and the constraint $\rho\gamma(\infty) < 1$ is known in this context as a small-loop-gain condition.

It is apparent, on examination of the argument given above, that we do not really need the restriction of $\Delta f(e)$ to hold for all e; if it holds only when e is within some fixed range, then the predicted bound on $|e - \hat{e}|$ will also continue to be valid so long as this condition remains satisfied, assuming it was initially so. Indeed, it may be quite adequate, in some cases, to take $f(e)$ as being actually linear when e is less than a certain amount, and then use the requirement that it should never exceed this value, as a design criterion; such systems have been called 'conditionally linear'. To exploit this idea, it is necessary to consider the structure of the feedback system in more detail, distinguishing between the plant, which is given at the outset, and the controller (or compensator), whose design is at our disposal. Moreover, we should also allow, in general, for separate nonlinearities in different parts of the loop, such as the actuator and sensor in Fig. 6.3, but the same methods can readily be extended to this case, and also to more general multivariable configurations.

An alternative approach for obtaining bounds on dynamical variables is through the use of Lyapunov functions. For simplicity, suppose we set $r = 0$ in Fig. 6.1, so that the operation of the feedback system is described by

$$\dot{\mathbf{x}} = \mathbf{A}_K\mathbf{x} + \mathbf{B}\{q + \Delta f(-\mathbf{C}\mathbf{x})\}$$

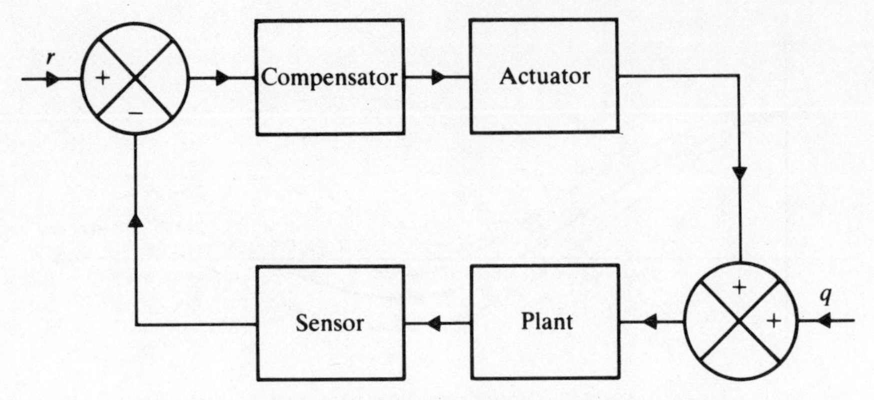

Fig. 6.3 Feedback loop with separated nonlinear elements.

where

$$\mathbf{A}_K = \mathbf{A} - \mathbf{B}K\mathbf{C}.$$

We can define a Lyapunov function

$$V = \mathbf{x}^{\mathrm{T}}\mathbf{P}\mathbf{x}$$

where the constant symmetric matrix $\mathbf{P}$ satisfies the Lyapunov matrix equation

$$\mathbf{P}\mathbf{A}_K + \mathbf{A}_K^{\mathrm{T}}\mathbf{P} = -\mathbf{I}$$

whose solution is positive-definite provided that $\mathbf{A}_K$ has all its eigenvalues in the open left half-plane, which is to say, the linearised closed-loop system is asymptotically stable. Then, we have

$$\begin{aligned} \dot{V} &= -\mathbf{x}^{\mathrm{T}}\mathbf{x} + 2\mathbf{x}^{\mathrm{T}}\mathbf{P}\mathbf{B}\{q + \Delta f(-\mathbf{C}\mathbf{x})\} \\ &\leqslant -|\mathbf{x}|^2 + 2|\mathbf{x}||\mathbf{P}\mathbf{B}|\{\mu + \rho|\mathbf{C}||\mathbf{x}|\} \end{aligned}$$

in view of the assumed constraints on q and $\Delta f(e)$, and it follows that $\dot{V} < 0$ whenever

$$|\mathbf{x}| > \frac{2\mu|\mathbf{P}\mathbf{B}|}{1 - 2\rho|\mathbf{P}\mathbf{B}||\mathbf{C}|}$$

assuming that the quantity on the right-hand side of the last inequality is positive. Now, since $\mathbf{P}$ is a symmetric, positive-definite matrix, it has a 'matrix square-root', that is, a matrix $\mathbf{R}$, also symmetric and positive-definite, such that

$$\mathbf{P} = \mathbf{R}^2$$

and so

$$V = |\mathbf{R}\mathbf{x}|^2.$$

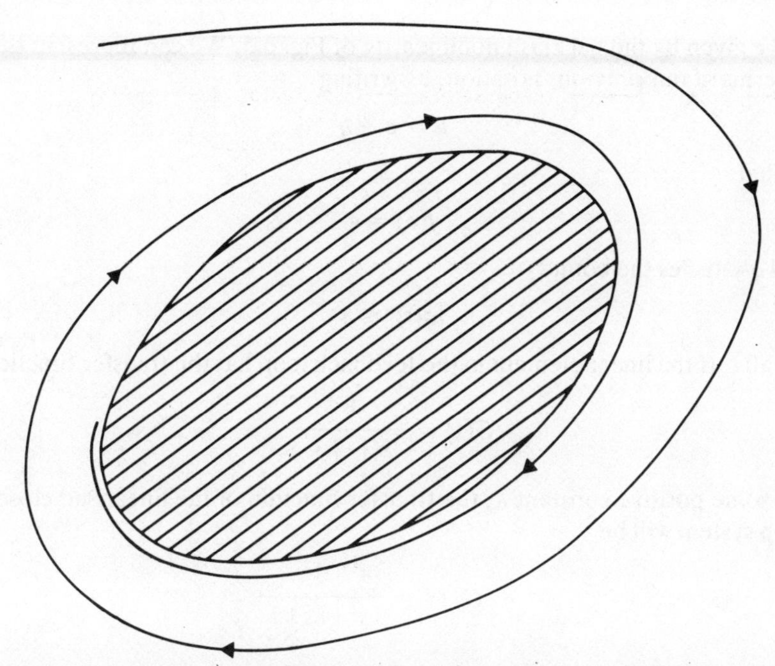

Fig. 6.4 Trajectory asymptotically approaching hypersurface of constant V.

Consequently, if $\Lambda(\mathbf{R})$ denotes the largest eigenvalue of $\mathbf{R}$, then V must always be decreasing when

$$|\mathbf{Rx}| > \frac{2\mu|\mathbf{PB}|\Lambda(\mathbf{R})}{1-2\rho|\mathbf{PB}|\,|\mathbf{C}|}$$

and hence $\mathbf{x}(t)$ eventually enters the region bounded by the corresponding hypersurface of constant V and remains there, or approaches it asymptotically as in Fig. 6.4. In terms of the error signal

$$e = -\mathbf{Cx} = -\mathbf{CR}^{-1}\mathbf{Rx}$$

this means that we ultimately have

$$|e(t)| \leqslant \frac{2\mu|\mathbf{PB}|\Lambda(\mathbf{R})|\mathbf{CR}^{-1}|}{1-2\rho|\mathbf{PB}||\mathbf{C}|}$$

in the sense that the upper limit of $|e(t)|$, as $t \to \infty$, satisfies this bound. Of course, the bounds obtained in this way will depend on the choice of Lyapunov function, and we could obtain an infinite number of different, but equally valid, results by making other choices.

Example 6.1 The effect of backlash

Let us consider again the system of Fig. 6.1, with the relation between u

and e given by the backlash nonlinearity of Fig. 6.5. We can represent this, in terms of our previous notation, by writing

$$u = e + q$$

so that

$$f(e) = e$$

and q satisfies the bound

$$|q(t)| \leq 1$$

for all t. If the linear element in the feedback loop has the transfer function

$$G(s) = \frac{1}{s(s+\lambda)}$$

for some positive constant λ, the transfer function of the linearised closed-loop system will be

$$G_K(s) = \frac{1}{s^2+\lambda s+1}$$

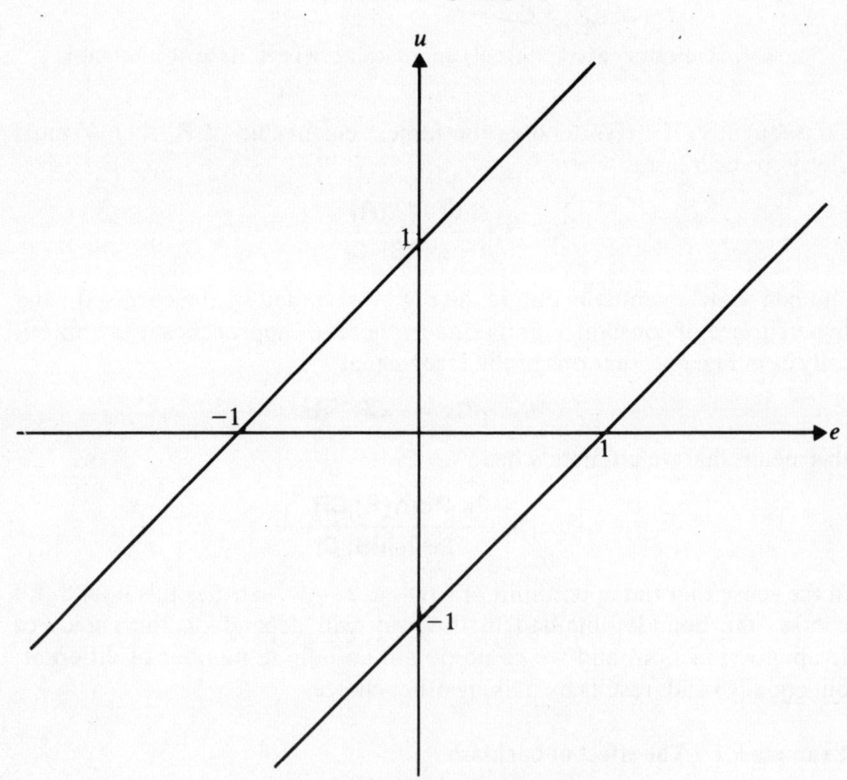

Fig. 6.5 Backlash nonlinearity.

since $K = 1$ in this case. Taking the inverse Laplace transform, we then get the impulse response function

$$g_K(t) = \exp\left(\frac{-\lambda t}{2}\right)\frac{\sin(\beta t)}{\beta}$$

where we have defined

$$\beta = \sqrt{(1 - \tfrac{1}{4}\lambda^2)}$$

assuming that $\lambda < 2$. The calculation of $\gamma(t)$ can now be done analytically, but the result is somewhat tedious to write down; however, letting $t \to \infty$, we get

$$\begin{aligned}
\gamma(\infty) &= \int_0^\infty \exp\left(\frac{-\lambda t}{2}\right)\frac{|\sin(\beta t)|}{\beta}\,\mathrm{d}t \\
&= \sum_{k=0}^{\infty} (-1)^k \int_{k\pi/\beta}^{(k+1)\pi/\beta} \exp\left(\frac{-\lambda t}{2}\right)\frac{\sin(\beta t)}{\beta}\,\mathrm{d}t \\
&= \sum_{k=0}^{\infty} \left(1 + \exp\left(\frac{-\lambda\pi}{2\beta}\right)\right)\exp\left(\frac{-k\lambda\pi}{2\beta}\right) \\
&= \frac{1 + \exp(-\lambda\pi/2\beta)}{1 - \exp(-\lambda\pi/2\beta)} \\
&= \coth\left(\frac{\lambda\pi}{4\beta}\right)
\end{aligned}$$

and so, since $\rho = 0$ and $\mu = 1$, we obtain the bound

$$|e(t) - \hat{e}(t)| \leqslant \coth\left(\frac{\lambda\pi}{4\beta}\right)$$

for all t. This result holds for $0 < \lambda < 2$; if, on the other hand, $\lambda \geqslant 2$, it turns out that the impulse response function is always non-negative, and hence

$$\gamma(\infty) = \int_0^\infty g_K(t)\,\mathrm{d}t = G_K(0) = 1$$

so that the bound on the error signal becomes

$$|e(t) - \hat{e}(t)| \leqslant 1$$

for all t. It is clear, on inspection of the backlash function, that this is the tightest bound we could hope to achieve, in general. Further, as might be expected, the bound (shown in Fig. 6.6) becomes looser as λ is decreased, making the system more resonant.

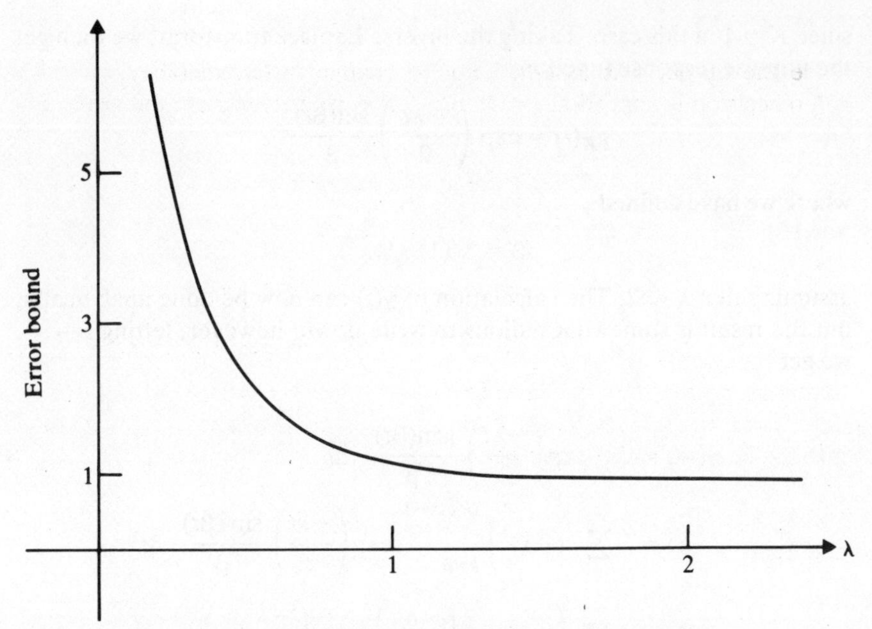

Fig. 6.6 Error bound for Example 6.1, as a function of λ.

6.2 The exponential-input describing function

When a stable control system is subjected to a sudden change in its reference input, it will normally settle to a new steady state, in a manner which can often be approximated by an exponential function of time. If the system is nonlinear, it then seems natural to analyse its behaviour approximately by using a modification of the describing function method, with an exponential, rather than sinusoidal, input. To illustrate this in the context of Fig. 6.1, let us take the disturbance $q = 0$ and assume that the nonlinear function $f(e)$ is single-valued. Further, we suppose that the system is initially in a steady state, with constant reference input $r = R_I$ and error signal $e = E_I$, when there is a step change to a new constant reference $r = R_0$, after which the error approaches its final steady value $e = E_0$. We thus have

$$E_I = R_I - G(0)\,f(E_I)$$
$$E_0 = R_0 - G(0)\,f(E_0)$$

and we then try to approximate the error signal, during the transitional stage, by writing

$$e(t) \simeq E_0 + E\exp(-\sigma t)$$

where σ is a positive constant, and

$$E = (E_I - R_I) - (E_0 - R_0)$$

since the output y is assumed not to change instantaneously when the reference step is applied at $t = 0$. Similarly, we approximate the signal $u = f(e)$ as

$$u(t) \simeq f(E_0) + U\exp(-\sigma t)$$

where

$$U = \int_0^\infty \{f(E_0 + E\exp(-\theta)) - f(E_0)\}\, d\theta$$

which represents an 'average' value of $\{f(e(t)) - f(E_0)\}$ over all positive t, in this approximation; the ratio U/E is then known as the exponential-input describing function (EIDF). Now, substituting the above approximants into the relation $e = r - y$, we get

$$E = -G(-\sigma)U$$

and hence

$$G(-\sigma)\frac{U}{E} = -1$$

which is analogous to the conditions for persistent oscillation obtained with the sinusoidal-input describing functions. Here, since E and E_0 are known, it can be used to estimate the rate σ at which $e(t)$ tends to its final value.

Example 6.2 Integral control system with saturation

If the transfer function $G(s)$ has a pole at $s = 0$, then we must have

$$f(E_I) = f(E_0) = 0$$

so that, if $f(e)$ vanishes only for $e = 0$, the steady-state error is always zero; such a system is said to have integral action. Suppose, in fact, that $f(e)$ represents the saturating nonlinearity of Fig. 6.7, with a piecewise-linear characteristic which makes it easy to calculate the EIDF. Since $E_0 = 0$, we have

$$U = \int_0^\infty f(E\exp(-\theta))\, d\theta = \int_0^E \frac{f(e)de}{e}$$

by a change of variable, so that $U = E$ for $|E| \leqslant 1$, and $|U| = 1 + \ln|E|$ for $|E| > 1$. Consequently, the EIDF is given by

$$\frac{U}{E} = \begin{cases} 1 & |E| \leqslant 1 \\ \dfrac{1 + \ln|E|}{|E|} & |E| > 1 \end{cases}$$

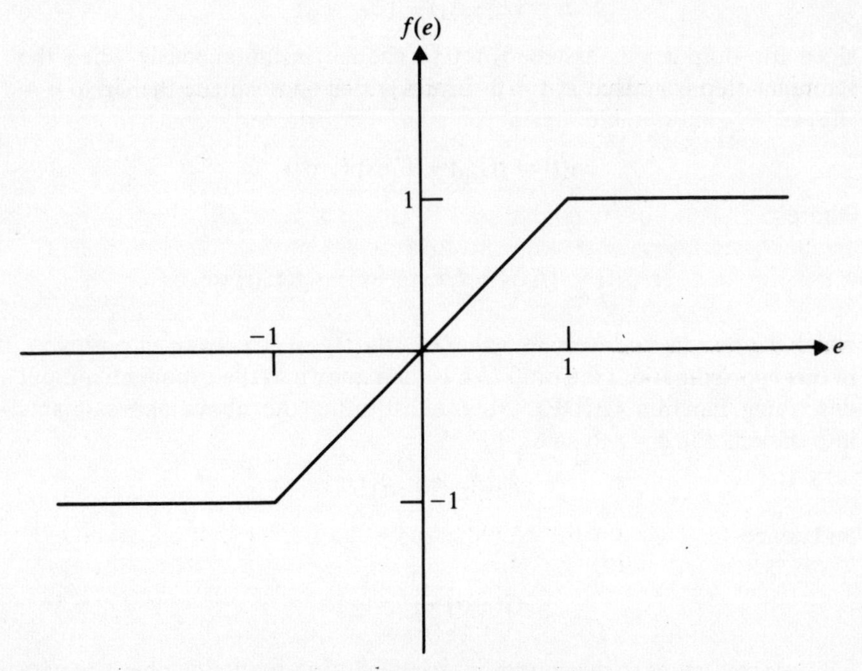

Fig. 6.7 Saturating nonlinearity.

as plotted in Fig. 6.8. Taking the simplest possible transfer function

$$G(s) = 1/s$$

we then get

$$\sigma = U/E$$

as an estimate of the 'mean' rate of approach to a steady state; it becomes slower for $|E| > 1$, because of the saturation.

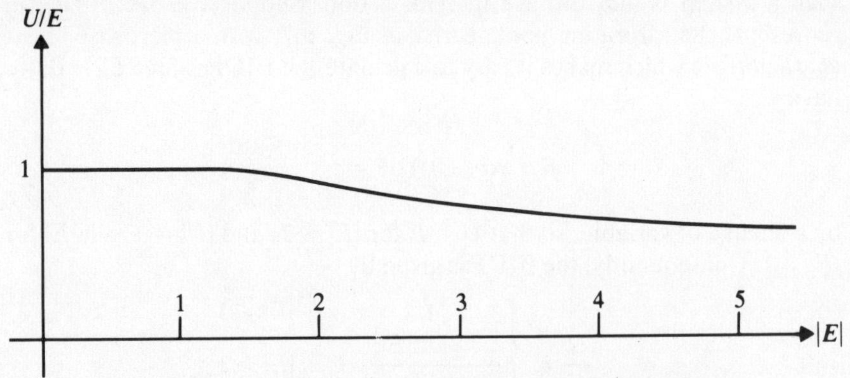

Fig. 6.8 Exponential-input describing function for Example 6.2.

6.3 Dither

Although nonlinear behaviour in a control system may be acceptable, it is, on the whole, undesirable since it can cause performance degradation and even instability. In order to alleviate these problems, it is sometimes practicable to make the nonlinear elements more approximately linear in effect, by causing the operating point to sweep repeatedly over a certain range around its nominal position, on a timescale much shorter than that of the system dynamics. This is done by injecting another external input, called a dither signal, into the system, just ahead of the nonlinearity; thus, in Fig. 6.1, we would add the dither $w(t)$ to the reference input, giving the arrangement shown in Fig. 6.9 where, to avoid irrelevant complications, we have removed the disturbance input. Writing $e = r-y$ as before, we now have

$$u = f(e + w)$$

where $w(t)$ is a rapidly oscillating function of time, which induces a kind of 'averaging' of the nonlinearity, depending on the nature of the oscillatory signal.

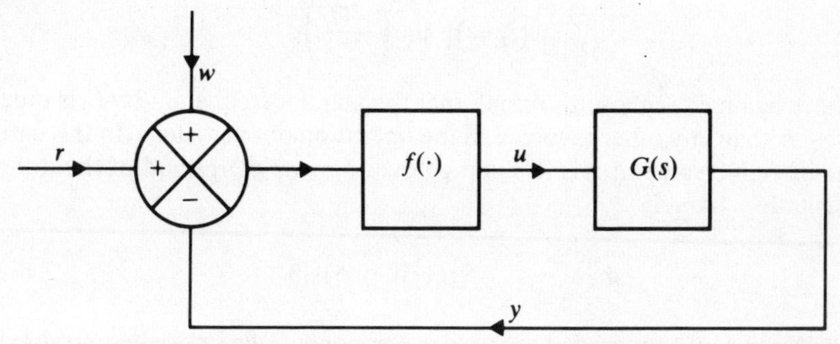

Fig. 6.9 Injection of dither signal into a feedback loop.

(i) **Square-wave dither**: The simplest type of dither signal is a square wave, as shown in Fig. 6.10, where $w(t)$ takes the constant values W and $-W$ alternately, each being held for a half-period $T_0/2$, with T_0 much smaller than the time constants of the system. As a result, the effective value of u, the output of the nonlinear element, can be written

$$\bar{u} = \frac{f(e+W) + f(e-W)}{2}$$

which is to say, the arithmetic mean of the values corresponding to the two levels of the dither signal.

(ii) **Sinusoidal dither:** Another simple possibility for the dither signal is a high-frequency sinusoid

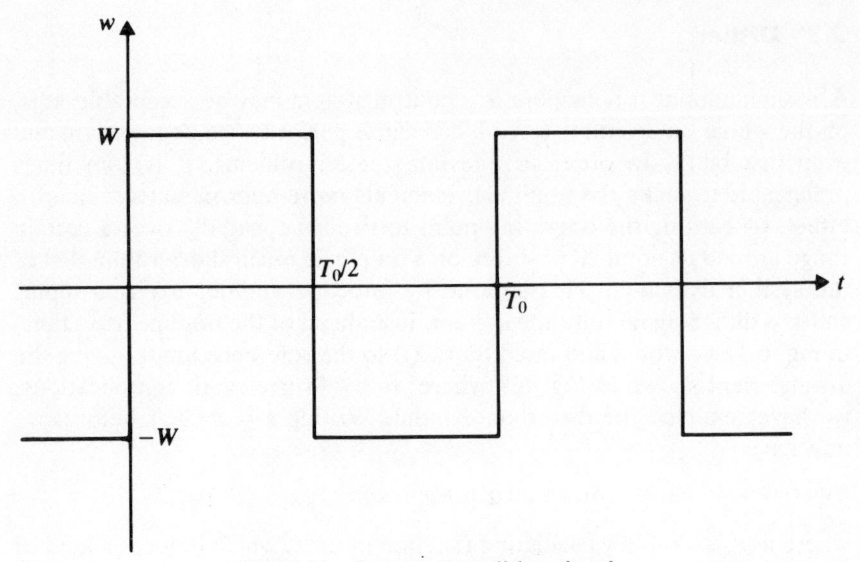

Fig. 6.10 Square-wave dither signal.

$$w(t) = W \sin\left(\frac{2\pi t}{T_0}\right)$$

where again T_0 is chosen so small that the angular frequency $2\pi/T_0$ is much greater than any other involved in the operation of the system. In this case, the effective value of u is its average over a complete period of the dither oscillation, namely

$$\bar{u} = \frac{1}{2\pi}\int_0^{2\pi} f(e + W\sin\theta)\,d\theta$$

which can also be regarded as the bias component of a Fourier expansion in multiples of the dither frequency. The approximation thus corresponds to the neglect of all harmonic terms, including the fundamental.

(iii) **Triangular-wave dither**: A third variety of dither, which is sometimes used, is the 'triangular' or 'saw-tooth' signal illustrated in Fig. 6.11, where $w(t)$ varies linearly with respect to time from $-W$ to W, and then back again. Because of the constancy of $\dot{w}$ on each upward or downward sweep, averaging over t is equivalent to averaging over w, and so the effective value of u becomes

$$\bar{u} = \frac{1}{2W}\int_{-W}^{W} f(e+w)\,dw$$

which is independent of the speed at which w actually changes. Thus, it is not in fact necessary that the times taken by the two linear sections of the

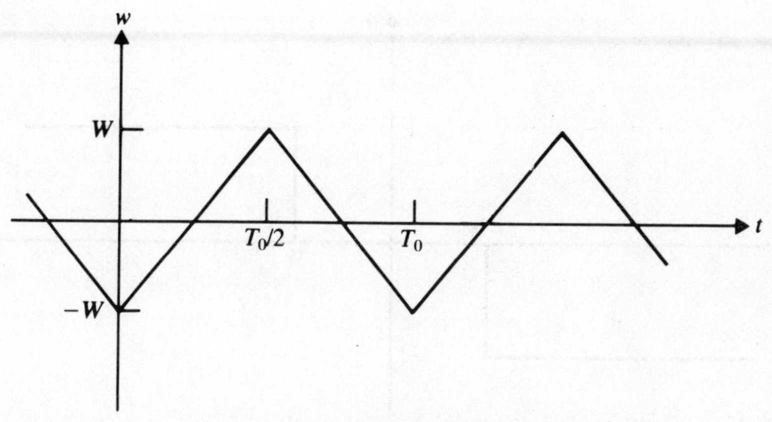

Fig. 6.11 Triangular-wave dither signal.

oscillation should be the same, provided that the total period T_0 is small enough, in the same sense as before. However, if $f(e)$ is multivalued, the upward and downward averages may be different, and then $\bar{u}$ is given by their weighted mean value, with weighting factors proportional to the durations of the corresponding sweeps.

Example 6.3 Smoothing of a discontinuous nonlinearity

For a comparison of the effects of different kinds of dither, let us take the ideal relay characteristic

$$f(e) = \mathrm{sgn}(e)$$

as in Fig. 6.12. With a dither signal covering only the range from $-W$ to W, with the constant $W > 0$, it is clear that we shall have

$$\bar{u} = \mathrm{sgn}(e)$$

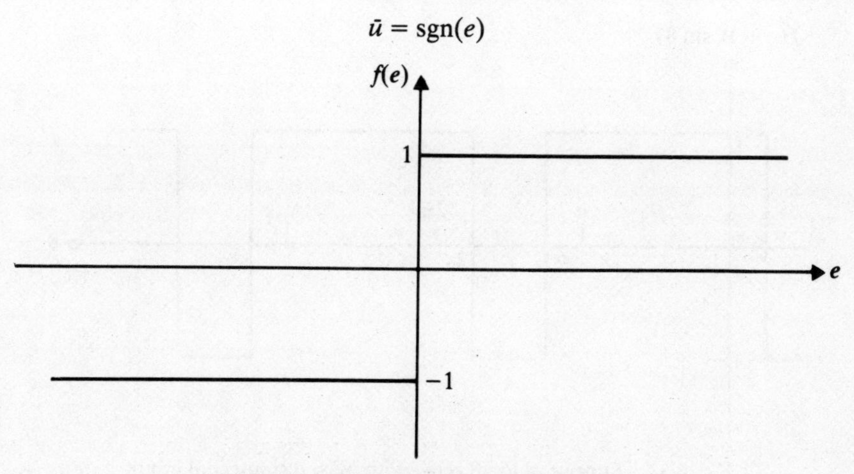

Fig. 6.12 Characteristic for ideal relay.

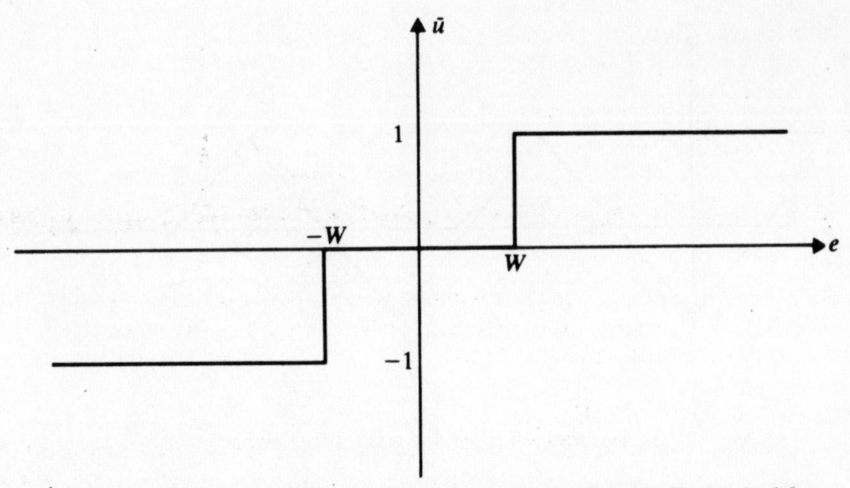

Fig. 6.13 Effective nonlinearity due to square-wave dither in Example 6.3.

whenever $|e| \geqslant W$, since u is then constant; thus, from here onwards, we take $|e| < W$. Then, with a square-wave dither, we get simply

$$\bar{u} = \frac{1 + (-1)}{2} = 0$$

and so the effect is to replace the original discontinuity in the nonlinear function by two discontinuities of half the size, as illustrated in Fig. 6.13; this may be an improvement, but does not constitute very effective smoothing.

Next, taking a sinusoidal dither signal instead, we have to integrate the function $f(e+W\sin\theta)$, shown in Fig. 6.14, which switches from -1 to 1

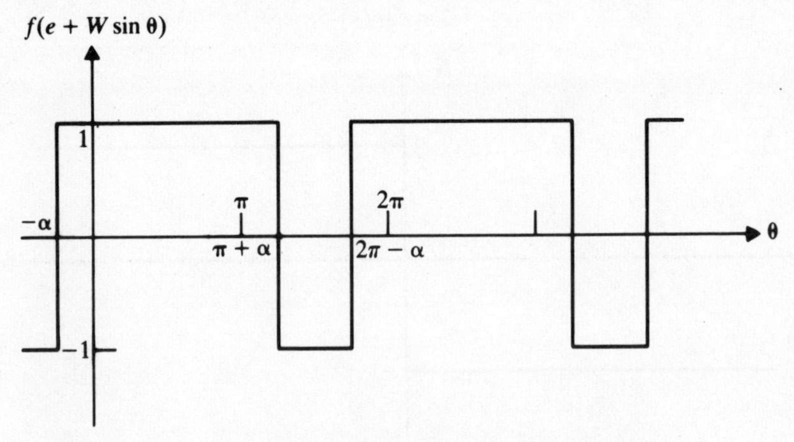

Fig. 6.14 Output of ideal relay with biased sinusoidal input.

at $\theta = -\alpha$, back to -1 at $\theta = \pi + \alpha$, back again at $\theta = 2\pi - \alpha$, and so on, where

$$\alpha = \arcsin(e/W).$$

Thus, re-specifying the integration range as being from $-\alpha$ to $2\pi - \alpha$, we get

$$\bar{u} = \frac{1}{2\pi}\left(\int_{-\alpha}^{\pi+\alpha} \mathrm{d}\theta - \int_{\pi+\alpha}^{2\pi-\alpha} \mathrm{d}\theta\right)$$
$$= \frac{2}{\pi}\arcsin(e/W)$$

which is plotted in Fig. 6.15. It is apparent that $\bar{u}$ is now a continuous function of e, but still one would hardly call it smooth, since it has infinite slope at the points where $|e| = W$.

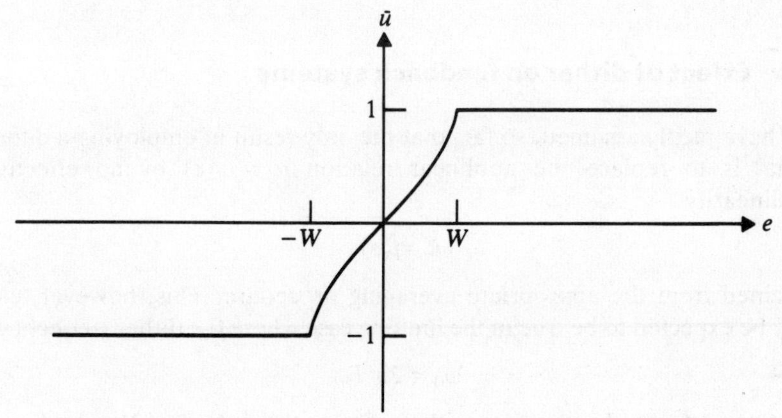

Fig. 6.15 Effective nonlinearity due to sinusoidal dither in Example 6.3.

Finally, if we use a triangular shape for the dither signal, the function to be integrated is $f(e+w)$, which changes from -1 to 1 when $w = -e$, giving

$$\bar{u} = \frac{1}{2W}\left(\int_{-e}^{W} \mathrm{d}w - \int_{-W}^{-e} \mathrm{d}w\right)$$
$$= e/W$$

so that the effective nonlinearity is now as shown in Fig. 6.16. This is evidently the smoothest of the three forms obtained, having the typical shape of a piecewise-linear saturation characteristic, and it is indeed normally true that the triangular dither produces a smoother effect than the others.

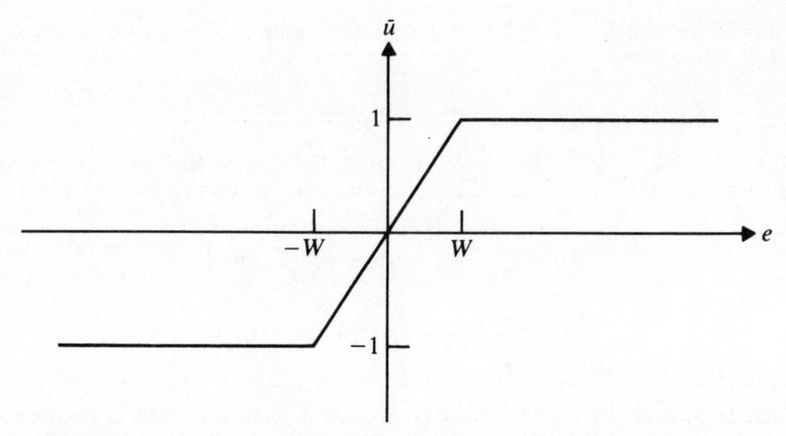

Fig. 6.16 Effective nonlinearity due to sawtooth dither in Example 6.3.

6.4 Effect of dither on feedback systems

We have tacitly assumed, so far, that the only result of employing a dither signal is to replace the nonlinear relation $u = f(e)$ by an effective nonlinearity

$$\bar{u} = \bar{f}(e)$$

obtained from the appropriate averaging procedure. This, however, can only be expected to be true in the limiting case where the dither frequency

$$\omega_0 = 2\pi/T_0$$

becomes infinite. In practice, with ω_0 finite, the output of the nonlinear element will contain an oscillation at this frequency and, for satisfactory operation of the control system, it will be necessary that this should be adequately suppressed by the dynamics of the feedback loop. For an approximate investigation of this process, we turn to harmonic analysis, which is most readily applicable in the case of sinusoidal dither, although it actually applies to the other kinds as well, since the square and triangular waves can be approximated by sinusoids of amplitude $4W/\pi$ and $8W/\pi^2$ respectively, when frequencies higher than the fundamental are ignored.

Assuming, then, that the dither is sinusoidal, we now approximate the signal $u(t)$ by the sum of the bias and fundamental-frequency terms in its Fourier expansion with respect to the dither period T_0, over which e is taken to be effectively constant. Now, this is the same approximation as was used, in Chapter 3, to construct the DIDF, so that we get

$$u(t) \simeq N_0(e,W)e + \operatorname{Im}\{N_1(e,W)W\exp(\mathrm{i}\omega_0 t)\}$$

where the first term on the right-hand side can immediately be identified with $\bar{u}$, since

$$N_0(e,W) = \frac{1}{2\pi e}\int_0^{2\pi} f(e+W\sin\theta)\,d\theta = \frac{\bar{u}}{e}$$

for a sinusoidal dither signal. Thus, in this approximation, we can write

$$u \simeq \bar{f}(e) + \Delta u$$

where

$$\Delta u(t) = \mathrm{Im}\{N_1(e,W)W\exp(i\omega_0 t)\}$$

and correspondingly, the output of the linear element can be written as

$$y \simeq \bar{y} + \Delta y$$

where

$$\Delta y(t) = \mathrm{Im}\{G(i\omega_0)N_1(e,W)W\exp(i\omega_0 t)\}$$

and $\bar{y}(t)$ is the signal which would appear at this point in the limit $\omega_0 \to \infty$. That is to say, $\bar{y}$ is the output which the linear element would produce when given the input $\bar{f}(r-\bar{y})$, so that it is determined by the equivalent feedback loop of Fig. 6.17. Of course, for a finite value of ω_0, the term Δy will be nonzero and will affect the error signal, making

$$e \simeq (r-\bar{y}) - \Delta y$$

but, if ω_0 is large enough for $|G(i\omega_0)N_1|$ to be very much less than unity, the extra term will be negligible compared with the dither signal which is being added to e in any case. Moreover, since we are assuming that $G(s) \to 0$ as $s \to \infty$, we can make the output perturbation Δy arbitrarily small by choosing a sufficiently high dither frequency. The overall effect of the dither is then simply the consequence of replacing $f(e)$ by $\bar{f}(e)$, which usually, though not always, improves the system's behaviour. For example, it is sometimes used to suppress limit cycles, but may also excite them instead, as can be seen from a further application of the describing function method, to the equivalent system of Fig. 6.17, in the following example.

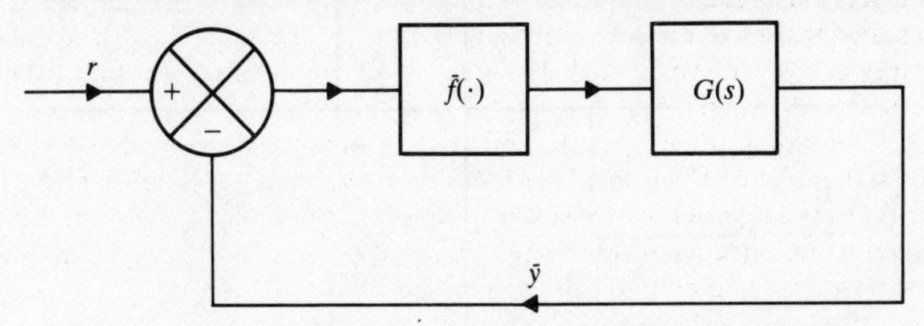

Fig. 6.17 Equivalent feedback loop in limit of infinite dither frequency.

Example 6.4 Ocillations under the influence of dither

With the ideal relay of Fig. 6.12 inserted into a feedback loop as in Fig. 6.9, the describing function method, in the absence of dither, would predict a limit cycle whenever the Nyquist plot of $G(s)$ crosses the negative real axis, since the SIDF takes all values from 0 to ∞. The effect of the dither, however, is to replace the nonlinearity by a function $\bar{f}(e)$, satisfying

$$0 \leqslant \frac{\bar{f}(e)}{e} \leqslant \frac{1}{W}$$

for $e \neq 0$, as found in Example 6.3, where W is the dither amplitude, so that the SIDF of $\bar{f}(e)$ must also lie between 0 and $1/W$. Consequently, the intersection condition for a limit cycle may well be no longer satisfied, in which case we predict, using the describing function approximation, that it has been suppressed by the dither signal. Indeed, even without the approximation, it would follow from the circle or Popov criterion that, provided the linear element is asymptotically stable, then so (for large enough W) is the system of Fig. 6.17, containing the effective nonlinearity.

On the other hand, suppose the original nonlinear function is a relay characteristic with a dead zone, so that

$$f(e) = \frac{\operatorname{sgn}(e+1) + \operatorname{sgn}(e-1)}{2}$$

as illustrated in Fig. 6.18. Since the effective nonlinearity $\bar{f}(e)$ is a linear functional of $f(e)$, it follows immediately from the results of Example 6.3 that we now have, in the case of sinusoidal dither,

$$\bar{f}(e) = \frac{1}{\pi}\left\{ \arcsin\left(\frac{e+1}{W}\right) + \arcsin\left(\frac{e-1}{W}\right)\right\}$$

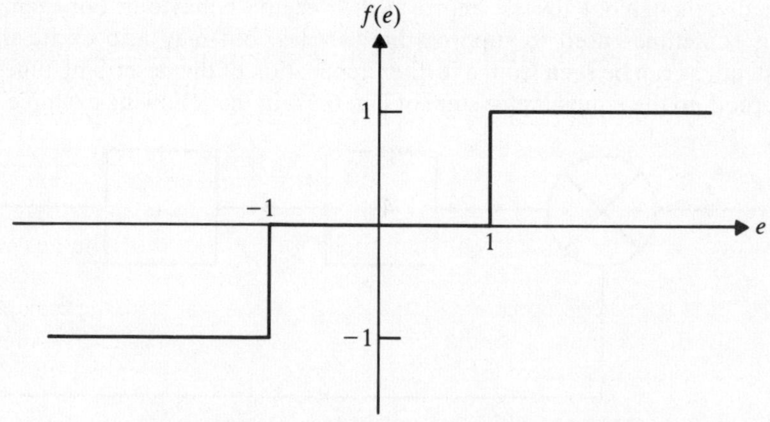

Fig. 6.18 Characteristic for relay with dead zone.

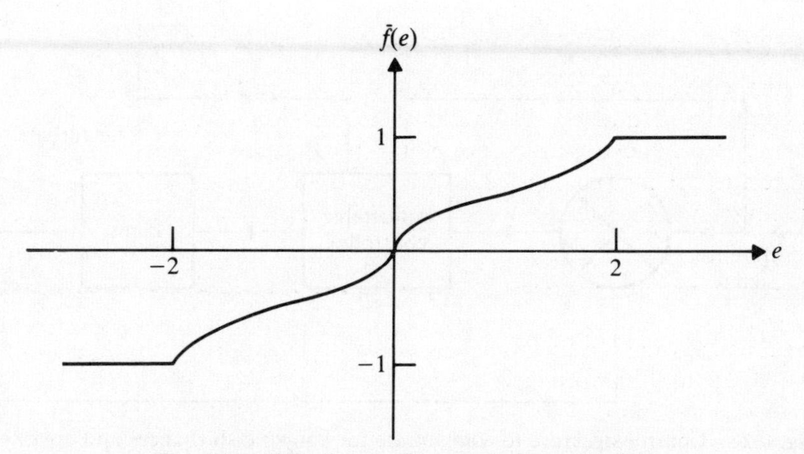

Fig. 6.19 Effective nonlinearity in Example 6.4.

with the understanding that the arcsine is to be replaced by the function πsgn/2 whenever its argument has modulus greater than unity. For the particular choice $W = 1$, the form of the function $\bar{f}(e)$ is as shown in Fig. 6.19, with infinite slope at the origin, so that its SIDF becomes unboundedly large as the input signal amplitude tends to zero. As a result, the limit cycle condition, given by the describing function analysis, can become satisfied for the system in Fig. 6.17, even if it was not so in the original dither-free system, with the implication that an oscillation has actually been induced by the dither.

6.5 Adaptive control

The most obvious difficulty presented by a nonlinear system, with regard to control, is that its behaviour depends, in general, on the operating conditions, so that a fixed linear controller may be inappropriate. A possible way of tackling this problem is then to incorporate some variable parameters into the control scheme, which can thus be continuously modified in order to deal with changing conditions or, in other words, can adapt to its environment. This type of control structure can, in fact, take several different forms, depending on how much is known about the dynamics of the plant, and what method is used to deal with the unknown aspects. If a reliable model is available, showing how the plant is affected by external changes, then the controller parameters can be made to depend explicitly on the reference input and environmental disturbances (assuming that these are measureable), in such a way as to compensate for their effects, giving a structure of the kind illustrated in Fig. 6.20.

More commonly, however, the need for adaptation arises in cases

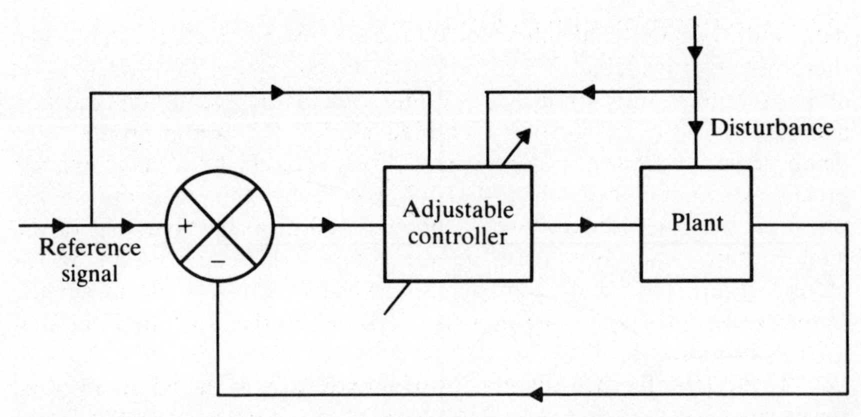

Fig. 6.20 Control structure to compensate for known disturbances and operating conditions.

where some aspects of the plant's behaviour are uncertain, inadequately modelled or subject to unpredictable changes in the course of time. There are then two main approaches which can be followed, and these are often referred to as 'direct' and 'indirect' adaptive control. In the indirect method, one takes the view that, if a linearised plant model, with known parameters, were available, a controller could then be designed automatically, for instance, by using one of the standard pole-assignment or optimisation algorithms provided by modern control theory. The adaptation therefore proceeds in two stages: first, a plant model is 'identified', that is to say, its parameters are estimated from measured input and output

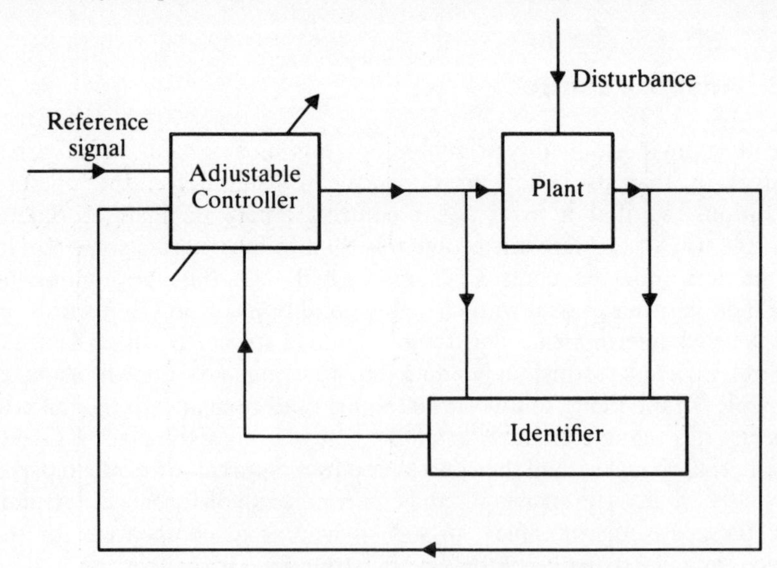

Fig. 6.21 Structure of a self-tuning control system.

data, and then the controller is 'updated' in accordance with the model thereby made available. These two steps are then repeated indefinitely, either simultaneously or in alternation, so that the controller is always adapted to the most recently identified model, the whole process now being generally known as 'self-tuning'. This indirect kind of adaptation procedure, schematically illustrated in Fig. 6.21, lies somewhat outside the range of this book, on account of the essentially stochastic nature of the identification stage. By contrast, the direct type, which involves no explicit identification, is usually formulated within a purely deterministic framework, the most common version being the so-called 'model-reference' approach.

In model-reference adaptive control (MRAC), as the name suggests, the basic idea is to compare the behaviour of the controlled plant with that of a reference model, representing the desired performance, and attempt to reduce the difference between them by changing the controller parameters in an appropriate way. The resulting structure is indicated in Fig. 6.22, and the essential requirement in designing such a scheme is to choose the adaptation law so as to ensure that the dynamical properties of the controlled system converge asymptotically to those of the reference model, which itself needs to be selected in such a way that this is achievable. Several different proposals for doing this have been made in recent decades, but none of them are entirely satisfactory and the subject is still in

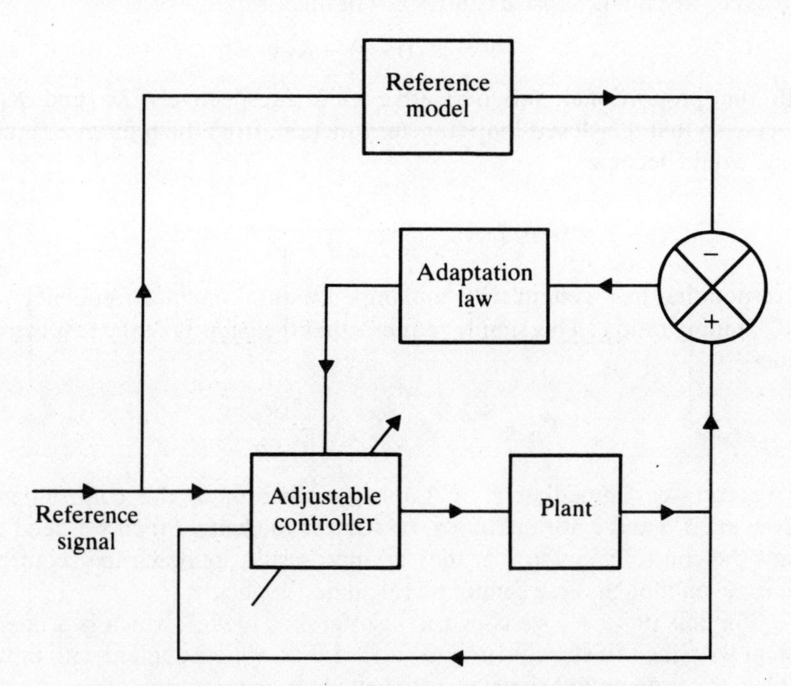

Fig. 6.22 Model-reference adaptive control system structure.

a state of development. Early heuristic suggestions, based on optimisation of an index of performance, were beset by stability problems, although these have since been overcome, to some extent, through the use of more systematic design procedures. In such methods, stability is built into the design by basing it on the existence of Lyapunov functions or the satisfaction of input-output criteria, though these considerations are somewhat restrictive and often lead to rather complicated structures. Typically, the property of strict positive-realness, as defined in Chapter 5, is required for the reference model, although it can sometimes be avoided by the introduction of auxiliary signals and/or derivative operators, which may, however, cause difficulties in the presence of noise. Moreover, the robustness of such systems against the effects of unmodelled dynamics in the plant is still an unresolved issue.

Example 6.5 A model-reference adaptive control system

Suppose the plant, with input u and output y, is described by a linearised model, having the transfer function

$$G(s) = \frac{b}{s(s+c)}$$

for some constants b and c. If the values of these parameters were known and fixed, we could choose a control law of the form

$$u = K_P(r-y) - K_D\dot{y}$$

with the proportional and derivative gains, respectively K_P and K_D, selected so that the closed-loop transfer function, from the reference signal r to y, would become

$$G_M(s) = \frac{\omega_n^2}{s^2 + 2\zeta\omega_n s + \omega_n^2}$$

corresponding to a system with undamped natural angular frequency ω_n and damping ratio ζ. This simply requires that the gains take the respective values

$$\hat{K}_P = \frac{\omega_n^2}{b}, \qquad \hat{K}_D = \frac{2\zeta\omega_n - c}{b}$$

as we can see immediately by Laplace-transforming the control law. However, if b and c are unknown, or subject to change, then we need to adapt the control gains so that they asymptotically approach the required values, even though these cannot be calculated explicitly.

For this purpose, we construct a reference model, which is a linear system with the transfer function $G_M(s)$, so that, when supplied with input r, it generates an output signal y_R satisfying

$$\ddot{y}_R + 2\zeta\omega_n \dot{y}_R = \omega_n^2 (r - y_R)$$

which can be compared with the equation

$$\ddot{y} + (bK_D + c)\dot{y} = bK_P (r - y)$$

satisfied by the plant output. Hence, defining the discrepancy between the plant and model outputs as an error signal

$$e_0 = y - y_R$$

we get

$$\ddot{e}_o + 2\zeta\omega_n \dot{e}_0 + \omega_n^2 e_o = b\{\phi_P(r-y) - \phi_D \dot{y}\}$$

where the deviations of the gains from their desired values have been denoted by

$$\phi_P = K_P - \hat{K}_P, \qquad \phi_D = K_D - \hat{K}_D,$$

respectively. We then try to choose the adaptation rules for the gains in such a way that there exists a Lyapunov function, whose properties will guarantee the stability of the system and also the convergence to zero of the gain deviations, as well as the output error signal.

From the infinite set of possible Lyapunov functions, the simplest form to select is

$$V = \dot{e}_0^2 + \omega_n^2 e_0^2 + \lambda_P \phi_P^2 + \lambda_D \phi_D^2$$

where λ_P and λ_D are arbitrary positive constants. Then, with the adaptation scheme

$$\dot{K}_P = \dot{\phi}_P = \mu_P (y - r)\, \dot{e}_0$$
$$\dot{K}_D = \dot{\phi}_D = \mu_D\, \dot{y}\, \dot{e}_0$$

it follows that

$$\dot{V} = -4\zeta\omega_n \dot{e}_0^2 \leqslant 0$$

provided that we take

$$\lambda_P = \frac{b}{\mu_P}, \qquad \lambda_D = \frac{b}{\mu_D},$$

which is always possible so long as μ_P and μ_D both have the same sign as b. It is thus necessary that the sign of b should be known, even if its value is not, but this will normally be true in practice. Although the result $\dot{V} \leqslant 0$ proves only the boundedness of the dynamical variables, convergence can also be established if the system is 'persistently excited' by the reference input, so that V cannot remain constant unless it vanishes. Incidentally, the appearance of the derivative signal $\dot{e}_0$ in the adaptation structure (shown in Fig. 6.23) is associated with the fact that $sG_M(s)$ is positive-real, though $G_M(s)$ is not.

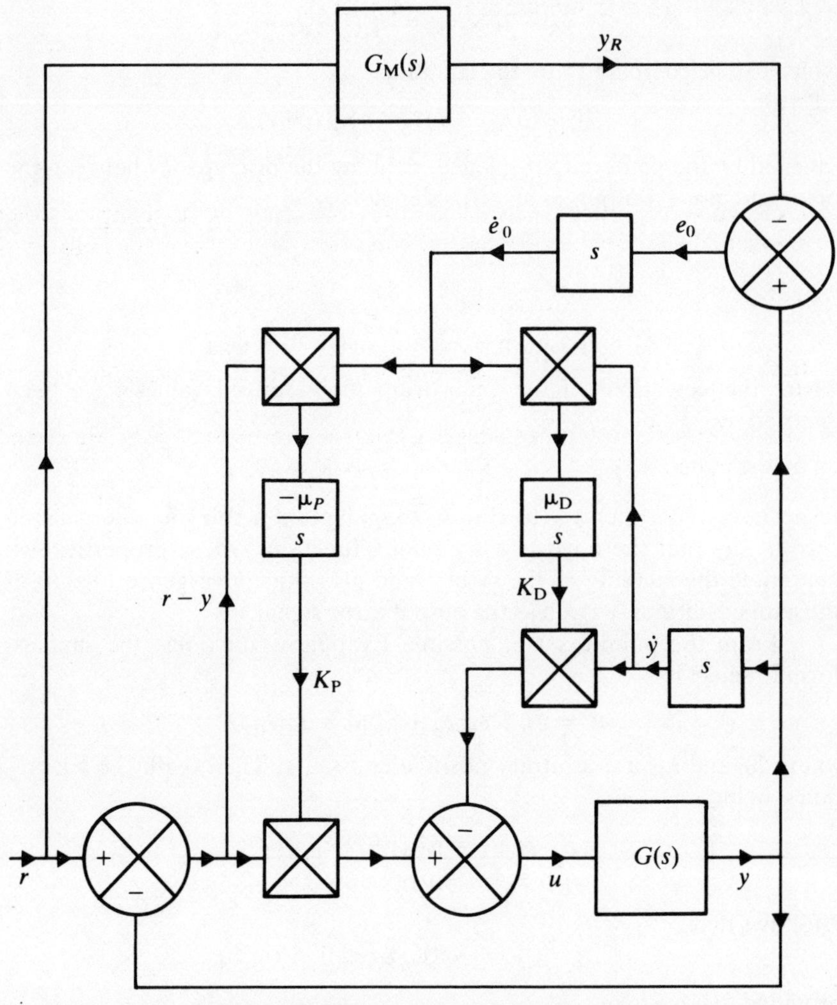

Fig. 6.23 MRAC system in Example 6.5.

6.6 Exercises

1 Calculate a bound on the error signal for a system as in Example 6.1, with the transfer function of the linear element replaced by

$$G(s) = \frac{2s+1}{s^2}.$$

2 Evaluate the EIDF for the cubic nonlinearity

$$u = e^3$$

with

$$e(t) = E_0 + E\exp(-\sigma t).$$

3 Determine the effective nonlinearities obtained by applying a dither signal at the input to a nonlinear element whose characteristic is given by

$$f(e) = \begin{cases} e & e \geqslant 0 \\ 0 & e < 0 \end{cases}$$

in the cases of square-wave, sinusoidal and triangular-wave dither.

4 Design a model-reference adaptive control system for a plant described by the equation

$$\dot{y} = au$$

where $a(>0)$ is fixed but unknown, with control law

$$u = K(r-y)$$

containing a variable gain K, and reference model transfer function

$$G_M(s) = \frac{1}{1+Ts}.$$

7 DISCRETE-TIME SYSTEMS

Throughout this book so far, we have been mainly concerned with systems defined in continuous time and governed by differential equations. However, especially in view of the widespread and increasing use of data sampling in modern technology, it seems appropriate, in this final chapter, to consider briefly the corresponding problems which arise if the time is discretised and the systems are described by recurrence relations. We have, in fact, already encountered such cases, particularly in connection with the point transformation method of Chapter 4, where the mapping between boundaries may be regarded in this way, and also when discussing linearisation around a limit cycle, in Chapter 5. Moreover, some dynamical models, notably in economics and biology, are more naturally formulated in discrete (rather than continuous) time, the discretisation being associated with, for example, the breeding cycle of an organism.

Let us then review the techniques applied in previous chapters to continuous-time systems, with the aim of extending their use, if possible, into the discrete-time area. The state-space formulation generalises very readily, as we saw in Chapter 1, with the time-derivatives in the equations being simply replaced by differencing or updating operations, but the concept of a state trajectory is no longer so useful, since the evolution is now represented by a sequence of points, instead of a continuous curve. As a result, the variety of dynamical behaviour is in some respects even greater for discrete-time systems, with chaotic motion, for instance, being possible even in a one-dimensional state-space. Somewhat similar considerations apply also to the method of harmonic balance, because again we do not have continuous functions of time, suitable for Fourier analysis. Nevertheless, with certain reservations, the describing function technique can be extended to sampled-data systems, albeit at the expense of considerable complications, as we shall see. On the other hand, the approximation of nonlinearities by piecewise-linear function is not very helpful with regard to discrete-time equations, as there are, in general, no boundary crossing points, since the trajectories are discontinuous, and so the point transformation method is inapplicable; similarly, Tsypkin's method suffers from the same difficulty, in a frequency-domain context. Despite the limitations of particular analytical approaches, however, many aspects of continuous-time theory do have discrete-time counterparts: thus, in the study of stability properties, Lyapunov's methods are still available, with appropriate modifications to the formalism, and so is the input–output interpretation. Moreover, we can again introduce various kinds of signal norm, analogous to the L_1-(integral of modulus), L_2- and L_∞-norms in continuous time, and the use of the supremum (L_∞) norm, in particular, enables us to define bounded-input bounded-output stability and obtain bounds on dynamical variables. Also, adaptive control strategies can be (and often are) formulated and/or implemented in a discrete-time framework.

7.1 Nonlinear recurrence relations

The state-space representation of a discrete-time dynamical system, with input vector $\mathbf{u}$ and output vector $\mathbf{y}$, has the general form

$$\begin{aligned} \mathbf{x}(t+1) &= \mathbf{f}(\mathbf{x}(t),\mathbf{u}(t),t) \\ \mathbf{y}(t) &= \mathbf{h}(\mathbf{x}(t),\mathbf{u}(t),t) \end{aligned}$$

where $\mathbf{x}$ is the state vector, and the time t takes integer values only. If the system is time-invariant, so that the functions $\mathbf{f}$ and $\mathbf{h}$ do not explicitly depend on t, and the input is held fixed, with

$$\mathbf{u}(t) = \hat{\mathbf{u}}$$

for all t, then the equilibrium points $\hat{\mathbf{x}}$ in the state-space are the solutions of the vector equation

$$\hat{\mathbf{x}} = \mathbf{f}(\hat{\mathbf{x}},\hat{\mathbf{u}})$$

with the corresponding output given by

$$\hat{\mathbf{y}} = \mathbf{h}(\hat{\mathbf{x}},\hat{\mathbf{u}}).$$

Linearising the equations around such a point, and using the prefix Δ to denote small deviations from equilibrium values, we obtain the linear discrete-time system

$$\begin{aligned} \Delta\mathbf{x}(t+1) &= \mathbf{A}\Delta\mathbf{x}(t) + \mathbf{B}\Delta\mathbf{u}(t) \\ \Delta\mathbf{y}(t) &= \mathbf{C}\Delta\mathbf{x}(t) + \mathbf{D}\Delta\mathbf{u}(t) \end{aligned}$$

where the elements of the constant matrices ($\mathbf{A}$,$\mathbf{B}$,$\mathbf{C}$,$\mathbf{D}$) are, as in Chapter 2, the partial derivatives of the nonlinear functions, evaluated at the equilibrium point. Here, however, the necessary and sufficient condition for asymptotic stability (of the linearised system) is that all the eigenvalues of the matrix $\mathbf{A}$ should lie strictly inside the unit circle, so that

$$\mathbf{A}^k \to 0$$

as $k \to \infty$, since the general solution of the vector recurrence relation for $\Delta\mathbf{x}$ is

$$\Delta\mathbf{x}(t) = \mathbf{A}^t \Delta\mathbf{x}(0) + \sum_{k=0}^{t-1} \mathbf{A}^{t-1-k}\, \mathbf{B}\Delta\mathbf{u}(k)$$

for all $t>0$.

Returning now to the original nonlinear system, but still taking it to be time-invariant and keeping the input fixed, we can suppress the dependence on $\mathbf{u}$, giving an autonomous relation of the form

$$\mathbf{x}(t+1) = \mathbf{f}(\mathbf{x}(t))$$

to be solved for $\mathbf{x}$. Defining a sequence of vector functions $\phi^{(k)}(\mathbf{x})$ by setting

$$\phi^{(k+1)}(\mathbf{x}) \equiv \mathbf{f}(\phi^{(k)}(\mathbf{x}))$$

for $k \geqslant 0$ with

$$\phi^{(0)}(\mathbf{x}) \equiv \mathbf{x}$$

we then have the formal solution

$$\mathbf{x}(t) = \phi^{(t)}(\mathbf{x}(0))$$

for all $t \geqslant 0$. If the function $\phi^{(t)}$ has a 'fixed point', that is to say, there is a vector $\bar{\mathbf{x}}$ satisfying (for some $t>0$)

$$\bar{\mathbf{x}} = \phi^{(t)}(\bar{\mathbf{x}})$$

then the solution with initial condition $\mathbf{x}(0) = \bar{\mathbf{x}}$ returns to its starting point after t steps, so the system is said to have a limit cycle of period t; clearly, with this definition, an equilibrium point, regarded as a degenerate limit cycle, has period 1. The stability of such a solution can be studied by writing

$$\mathbf{x} = \bar{\mathbf{x}} + \tilde{\mathbf{x}}$$

where $\tilde{\mathbf{x}}$ denotes a small departure from the nominal trajectory, for which we immediately derive the linearised relation

$$\tilde{\mathbf{x}}(t) = \nabla_{\mathbf{x}}\phi^{(t)}(\bar{\mathbf{x}})\, \tilde{\mathbf{x}}(0)$$

in the case of a limit cycle with period t. From the definition of the functions $\phi^{(k)}$ as iterates of $\mathbf{f}$, the coefficient matrix here can be written as

$$\nabla_{\mathbf{x}}\phi^{(t)}(\bar{\mathbf{x}}) = \prod_{k=0}^{t-1} \nabla_{\mathbf{x}}\mathbf{f}(\phi^{(k)}(\bar{\mathbf{x}}))$$

and the limit cycle is stable if every eigenvalue of this matrix has modulus less than unity.

Example 7.1 A random number generator

Consider the simple one-dimensional recurrence relation

$$x(t+1) = 2|x(t)|-1$$

where x is a scalar variable. The solution is given by the iterates $\phi^{(t)}$ of the function

$$f(x) = 2|x|-1$$

which is shown in Fig. 7.1. Evidently, if the initial state satisfies the condition

$$|x(0)| \leqslant 1$$

then we have

$$|x(t)| \leqslant 1$$

for all $t>0$. Further, by considering the form of the functions $\phi^{(t)}$, it can be

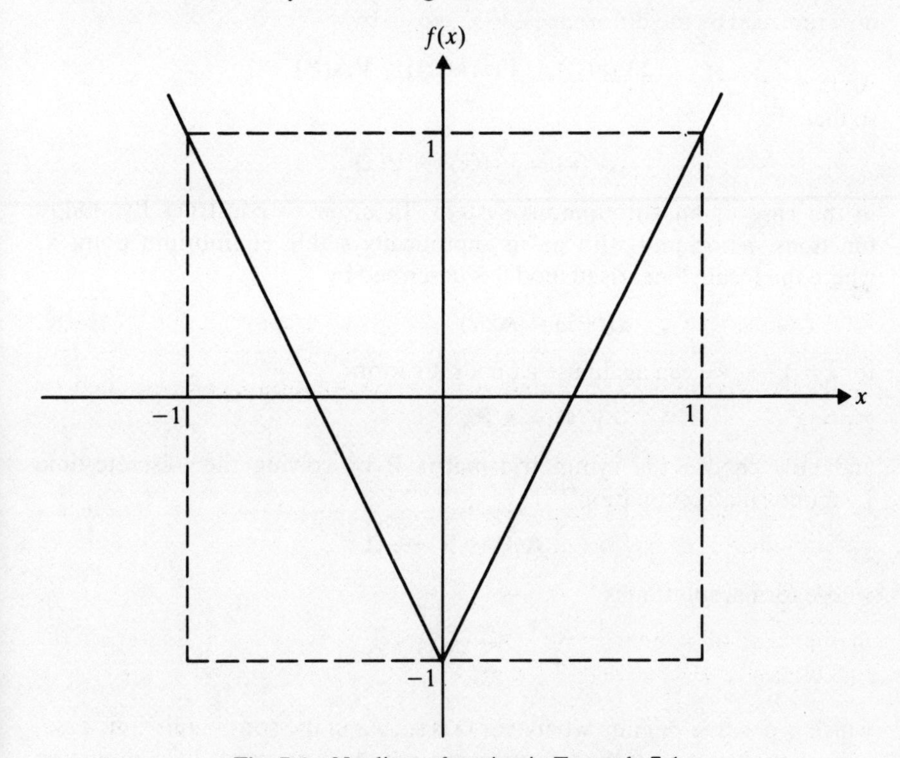

Fig. 7.1 Nonlinear function in Example 7.1.

seen that the system has limit cycles of every possible period, including the two equilibrium points $x = 1$ and $x = -1/3$. All these solutions, however, are unstable, since, if $\bar{x}$ is the starting point of a limit cycle having period t, then

$$|\nabla_x \phi^{(t)}(\bar{x})| = 2^t > 1.$$

Moreover, for almost all initial points (in fact, for all irrational ones) satisfying $|x(0)|<1$, the solution consists of a nonrepeating sequence, randomly distributed on the interval $(-1,1)$.

7.2 Stability in discrete time

Most of the concepts and techniques associated with stability in continuous time have fairly direct analogies for discrete-time systems, and here we shall simply indicate some of the main alterations which arise. We have already seen that local stability properties can be analysed by linearisation, which is to say, by Lyapunov's first method; similarly, we can study nonlocal properties by using the second method of Lyapunov, as in Chapter 5, except that the time-derivative of the scalar function $V(\mathbf{x})$ is now replaced by the difference $\Delta V(\mathbf{x})$, given by

$$\Delta V(\mathbf{x}(t)) = V(\mathbf{x}(t+1)) - V(\mathbf{x}(t))$$

so that

$$\Delta V(\mathbf{x}) = V(\mathbf{f}(\mathbf{x})) - V(\mathbf{x})$$

in the case of an autonomous system. In order to construct Lyapunov functions, associated with an asymptotically stable equilibrium point $\hat{\mathbf{x}}$, where the locally linearised model is described by

$$\tilde{\mathbf{x}}(t+1) = \mathbf{A}\tilde{\mathbf{x}}(t)$$

for $\tilde{\mathbf{x}} = \mathbf{x}-\hat{\mathbf{x}}$, we can again use a quadratic form

$$V(\mathbf{x}) = \tilde{\mathbf{x}}^T \mathbf{P} \tilde{\mathbf{x}}$$

and now choose the symmetric matrix $\mathbf{P}$ by solving the 'discrete-time Lyapunov matrix equation'

$$\mathbf{A}^T\mathbf{P}\mathbf{A} - \mathbf{P} = -\mathbf{Q}$$

whose formal solution is

$$\mathbf{P} = \sum_{k=0}^{\infty} (\mathbf{A}^T)^k \mathbf{Q} \mathbf{A}^k$$

which is positive-definite whenever $\mathbf{Q}$ is so. As in the continuous-time case, a Lyapunov function thus constructed can then be used, in conjunction with the nonlinear state-space equations, to estimate a domain of attraction. Extensions of Lyapunov's second method can further be applied to the investigation of unstable equilibria, stability of trajectories and also time-varying systems, described in the discrete-time context by equations of the form

$$\mathbf{x}(t+1) = \mathbf{f}(\mathbf{x}(t),t)$$

for which the Lyapunov functions will, in general, also be explicitly time-dependent, with

$$\Delta V(\mathbf{x},t) = V((\mathbf{f}(\mathbf{x}),t),t) - V(\mathbf{x},t).$$

The representation of an arbitrary dynamical system as a feedback loop comprising two blocks, one of which is linear and time-invariant, is also applicable in the discrete-time case, whether or not the original equations fall naturally into this form. Let us suppose, for simplicity, that there is only a single nonlinear relation involved, namely

$$u(t) = -F(\mathbf{x},t)\, y(t)$$

between the scalar variables u and y, which are also related by the state-space equations

$$\begin{aligned} \mathbf{x}(t+1) &= \mathbf{A}\mathbf{x}(t) + \mathbf{B}u(t) \\ y(t) &= \mathbf{C}\mathbf{x}(t) + \mathbf{D}u(t) \end{aligned}$$

with constant matrices $(\mathbf{A},\mathbf{B},\mathbf{C},\mathbf{D})$. We can then characterise the linear time-invariant block by a discrete-time transfer function

$$G(z) = \mathbf{C}(z\mathbf{I}-\mathbf{A})^{-1}\,\mathbf{B} + \mathbf{D}$$

for which a discrete-time version of the positive-realness concept can be introduced. In fact, we say that $G(z)$ is positive-real (in the discrete-time sense) if the function

$$G\left(\frac{1+s}{1-s}\right) = (1-s)\,\mathbf{C}\{(\mathbf{A}+\mathbf{I})s - (\mathbf{A}-\mathbf{I})\}^{-1}\,\mathbf{B} + \mathbf{D}$$

is positive-real in the previous (continuous-time) sense; also, strict positive-realness is similarly defined. Here, the significance of the transformation

$$z = \frac{1+s}{1-s}$$

is that it maps the left half-plane and imaginary axis of the s-plane into the unit disc and its boundary in the z-plane, these being the regions associated with stability in continuous and discrete time, respectively. Then, if

$$F(\mathbf{x},t) \geqslant 0$$

for all $\mathbf{x}$ and t, and the discrete-time positive-realness condition holds for $G(z)$, a Lyapunov function can be constructed, with which to prove the stability of the feedback system, as in the corresponding continuous-time problem; moreover, if $G(z)$ is, in the discrete-time sense, strictly positive-real, then global asymptotic stability can be established as well. Further, these results also admit a kind of frequency-domain interpretation, analogous to the circle criterion, since the required conditions on $G(z)$ can be checked from the graph of $G(\exp(i\theta))$ for real θ.

Example 7.2 A multiplicative recurrence relation

The second-order system described by the equations

$$x_1(t+1) = x_2(t)$$
$$x_2(t+1) = x_1(t)\, x_2(t)$$

clearly has two equilibrium points, at $x_1 = x_2 = 0$ and $x_1 = x_2 = 1$, which are (locally) stable and unstable, respectively, as is readily shown by linearisation. Using the Lyapunov function

$$V(\mathbf{x}) = x_1^2 + x_2^2$$

we find that

$$\Delta V(\mathbf{x}) = (x_2^2 - 1)\, x_1^2$$

and so $\Delta V(\mathbf{x}) \leqslant 0$ for $|x_2| \leqslant 1$; also, we can only have $\Delta V = 0$ with $|x_2| \neq 1$ if $x_1 = 0$. Hence, any solution which has its initial point in the region

$$x_1^2 + x_2^2 < 1$$

must approach the stable equilibrium ($\mathbf{x}=0$) asymptotically, unless it begins with $x_1 = 0$ or $x_2 = 0$, in which case it reaches the origin in a finite time and remains there. However, the system is not globally stable, since every solution with initial conditions in $(x_1>1, x_2>1)$ diverges to infinity.

Example 7.3 System with a quotient nonlinearity

This kind of structure can arise in several areas of control theory, the present case being taken from a self-tuning scheme. The state-space equations, in their autonomous form, can be written as

$$x_1(t+1) = \frac{x_1(t)}{x_2(t)}$$

$$x_2(t+1) = \rho\, \frac{x_1(t)}{x_2(t)} + (1-\rho)\, x_2(t)$$

where ρ is a constant between 0 and 1. Linearising around the equilibrium point at $x_1 = x_2 = 1$, we obtain

$$\mathbf{A} = \begin{pmatrix} 1 & -1 \\ \rho & 1-2\rho \end{pmatrix}$$

for the discrete-time evolution matrix, giving

$$\det(z\mathbf{I}-\mathbf{A}) = z^2-2(1-\rho)z + (1-\rho)$$

so that the eigenvalues of **A** lie inside the unit circle for all ρ with

$$0 < \rho < 1.$$

Consequently, the equilibrium is locally asymptotically stable; moreover, it appears from numerical studies that all solutions starting from points in the positive quadrant ($x_1>0$, $x_2>0$) converge to the equilibrium, though it is not clear how this can be rigorously proved.

7.3 Sampled-data systems

In control engineering, the principal applications of discrete-time theory are to the analysis and design of sampled-data control systems. The nonlinear aspects of this topic can be illustrated by considering the feedback system in Fig. 7.2, where the error signal $e(t)$, which is the difference between the reference input $r(t)$ and the plant output $y(t)$, is sampled at intervals of T_s, starting at $t = 0$, to give

$$e^*(t) = \sum_{k=0}^{\infty} e(kT_s)\delta(t-kT_s)$$

assuming that the sampling is instantaneous, so that it generates a train of delta-function impulses. Taking the Laplace transformation $e^*(t)$, we get

$$E^*(s) = \int_0^{\infty} e^*(t)\exp(-st)\mathrm{d}t = \sum_{k=0}^{\infty} e(kT_s)z^{-k}$$

where we have defined

$$z \equiv \exp(T_s s).$$

At this point, the sequence of coefficients $\{e(kT_s)\}$ is processed by a digital computer to give another sequence $\{u(kT_s)\}$, which may similarly be regarded as the coefficients of an impulse train

$$u^*(t) = \sum_{k=0}^{\infty} u(kT_s)\delta(t-kT_s)$$

with Laplace transform

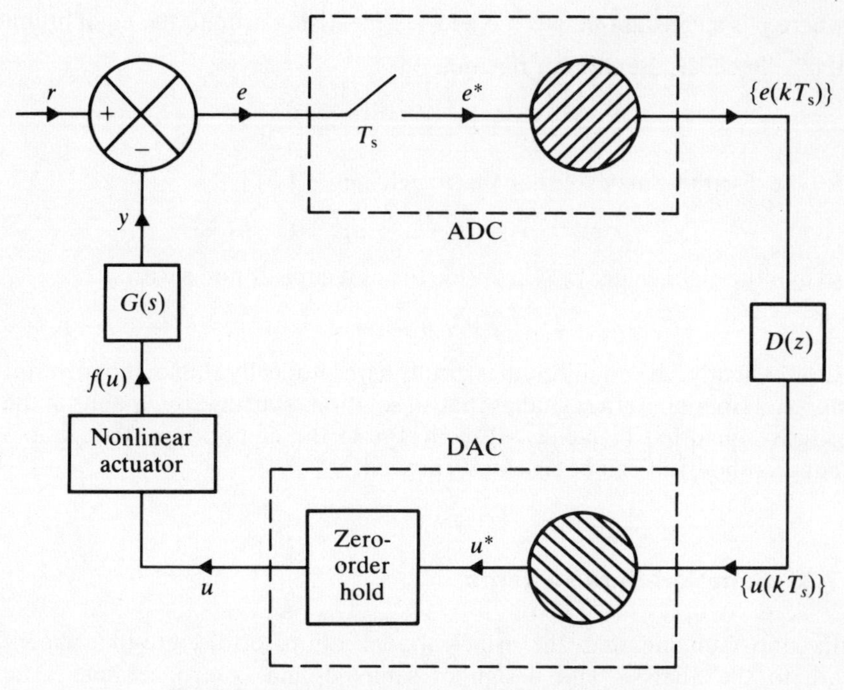

Fig. 7.2 A sampled-data feedback control system.

$$U^*(s) = \sum_{k=0}^{\infty} u(kT_s)z^{-k}.$$

If, as is usual, the purpose of the digital processor is to implement a linear control law, the coefficients in e^* and u^* will be related by an equation of the form

$$u(kT_s) = \sum_{j=0}^{k} d_{k-j}\, e(jT_s)$$

for some sequence of constants $\{d_k\}$, so that

$$U^*(s) = D(z)\, E^*(s)$$

if we define the 'discrete-time transfer function'

$$D(z) = \sum_{k=0}^{\infty} d_k z^{-k}$$

to characterise the operation. Actually, the generation of the sequence $\{e(kT_s)\}$, from the original signal $e(t)$, is carried out by a device called an analogue-to-digital converter (ADC) and, when the sequence $\{u(kT_s)\}$ is produced, the converse process of generating the control signal $u(t)$ is performed by a digital-to-analogue converter (DAC). Mathematically, this procedure is equivalent to acting on the signal $u^*(t)$ with a 'zero-order hold', which has the transfer function

$$\frac{1-z^{-1}}{s}=\frac{1-\exp(-T_s s)}{s}$$

and thus produces

$$\begin{aligned} U(t) &= \int_{t-T_s}^{t} u^*(\tau)\,d\tau \\ &= u(kT_s), \qquad kT_s<t<(k+1)T_s, \end{aligned}$$

so that the control signal is piecewise-constant. We then suppose that this signal is applied to a nonlinear actuator with characteristic $f(u)$, so that the state-space equations for the plant, which is assumed to be linear, can be written

$$\begin{aligned} \dot{\mathbf{x}} &= \mathbf{A}\mathbf{x} + \mathbf{B}f(u) \\ y &= \mathbf{C}\mathbf{x} \end{aligned}$$

where $\mathbf{x}$ is its state vector. Since the plant input $f(u(t))$ is piecewise-constant, we can solve the state-space equations to obtain

$$y(kT_s) = \mathbf{C}\exp(k\mathbf{A}T_s)\,\mathbf{x}(0) + \sum_{j=0}^{k-1} \mathbf{C}\exp\{(k-1-j)\,\mathbf{A}T_s\}\,\mathbf{B}_0 f(u(jT_s))$$

at the sampling instants, with

$$\mathbf{B}_0 = \mathbf{A}^{-1}\{\exp(\mathbf{A}T_s) - \mathbf{I}\}\mathbf{B}$$

provided that $\mathbf{A}$ is nonsingular; in general,

$$\mathbf{B}_0 = \int_0^{T_s} \exp(\mathbf{A}t)\,\mathbf{B}\,dt.$$

Thus, as the sampled values of the error are given by

$$e(kT_s) = r(kT_s) - y(kT_s)$$

we now have a complete set of discrete-time relations to determine the behaviour of the closed-loop system.

The stability properties of the system can then be investigated by using the discrete-time equations relating the sampled values of the signals, and also, provided that we are not concerned about inter-sample behaviour, the system's performance can be assessed in the same way. Eliminating e and y, and defining a sequence of coefficients

$$\alpha_k = \sum_{j=1}^{k} d_{k-j}\,\mathbf{C}\Phi^{j-1}\,\mathbf{B}_0$$

where

$$\Phi = \exp(\mathbf{A}T_s)$$

we obtain the equation

$$u(kT_s) = \sum_{j=0}^{k} d_{k-j}\{r(jT_s) - \mathbf{C}\Phi^j\mathbf{x}(0)\} - \sum_{j=0}^{k-1} \alpha_{k-j} f(u(jT_s))$$

to determine u at the sampling instants. Then, if we write

$$f(u) = Ku + \Delta f(u)$$

for some constant K, and define $\hat{u}(t)$ to be the value which $u(t)$ would have for $\Delta f \equiv 0$, it follows that

$$\tilde{u}(kT_s) = -\sum_{j=0}^{k-1} \alpha_{k-j} \{K\tilde{u}(jT_s) + \Delta f(u(jT_s))\}$$

with $\tilde{u} = u - \hat{u}$, whence bounds arising from restrictions on Δf can be derived, as for continuous systems in Chapter 6. In order to do this, we first define another sequence β_k from the coefficients of z^{-1} in the expression

$$\sum_{k=1}^{\infty} \beta_k z^{-k} = \frac{1}{K + \left(\sum_{k=1}^{\infty} \alpha_k z^{-k}\right)^{-1}}$$

which constitutes the closed-loop discrete-time transfer function relating $\Delta f(u)$ to $-\tilde{u}$, so that

$$\tilde{u}(kT_s) = -\sum_{j=0}^{k-1} \beta_{k-j} \Delta f(u(jT_s)).$$

Hence, if $\Delta f(u)$ can be written as

$$\Delta f(u) = \tilde{f}(u) + q(u)$$

with

$$|\tilde{f}(u)| \leqslant \rho|u|$$

and

$$|q| \leqslant \mu$$

for some constants μ and ρ, it follows that

$$|\tilde{u}(kT_s)| \leqslant \gamma_k \{\mu + \rho \max_{0 \leqslant j < k} |u(jT_s)|\}$$

where

$$\gamma_k = \sum_{j=1}^{k} |\beta_j|$$

and hence

$$|\tilde{u}(kT_s)| \leqslant \frac{\gamma_k}{1-\rho\gamma_k} \{\mu + \rho \max_{0 \leqslant j < k} |\hat{u}(jT_s)|\}$$

so long as

$$\gamma_k < 1/\rho.$$

We thus have an explicit bound on the deviation of u from $\hat{u}$, not only at the sampling instants but for all $t \geqslant 0$, as illustrated in Fig. 7.3.

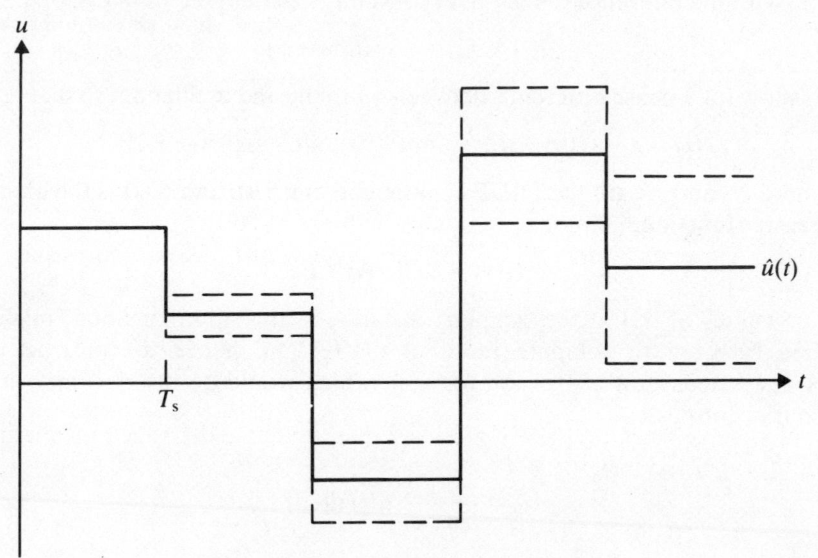

Fig. 7.3 Bounds on the control signal.

7.4 Limit cycles in sampled systems

Just as for continuous-time feedback systems, so also with sampled data, a typical manifestation of nonlinear behaviour is the appearance of limit cycles. Now, however, the situation is more complicated, because there are two periodicities involved, one due to the sampling and the other to the limit cycle oscillation. Nevertheless, it commonly happens, if the signals are not sampled too frequently, that the limit cycle period T turns out to be an exact multiple of the sampling period T_s, so that

$$T = mT_s$$

for some integer $m > 1$; further, if the nonlinearity has odd symmetry and either the system has integral action or the reference input is zero, we can expect m to be even, though there is no need to assume this. The fact that the angular frequency ω of the limit cycle is then a subharmonic of the sampling frequency, that is to say,

$$\omega = \frac{2\pi}{T} = \frac{2\pi}{mT_s}$$

makes the analysis much simpler, whether we use the describing function or other methods, so we will now confine our attention to this case.

Applying the describing function approximation to the system of Fig. 7.2, with the reference signal r held constant, we could set, as in Chapter 3,

$$u(t) \simeq U_0 + U_1 \sin(\omega t+\psi)$$

to allow for a phase difference between sampling and oscillation, so that

$$e(t) \simeq r - G(0)\, N_0 U_0 - \mathrm{Im}[G(\mathrm{i}\omega)\, N_1 U_1 \exp\{\mathrm{i}(\omega t+\psi)\}]$$

where N_0 and N_1 are the DIDF components for $f(u)$, and $G(s)$ is the plant transfer function,

$$G(s) = \mathbf{C}(s\mathbf{I}-\mathbf{A})^{-1}\,\mathbf{B}.$$

The values of $e(t)$ at the sampling instants, in this approximation, might then be used to compute those of $u(kT_s)$ and hence to construct a piecewise-constant expression for $u(t)$, which would be inserted into the Fourier formulae

$$U_0 = \frac{1}{2\pi}\int_0^{2\pi} u(t)\, \mathrm{d}(\omega t)$$

$$U_1 \cos\psi = \frac{1}{\pi}\int_0^{2\pi} u(t) \sin(\omega t)\, \mathrm{d}(\omega t)$$

$$U_1 \sin\psi = \frac{1}{\pi}\int_0^{2\pi} u(t) \cos(\omega t)\, \mathrm{d}(\omega t)$$

giving three equations to be solved for (U_0, U_1, ψ), these being the only unknown quantities since ω is already determined when the integer m is selected. On the other hand, the validity of the DIDF approximation here is somewhat dubious, in view of the fact that we are dealing with a subharmonic oscillation, so that a multiple-input describing function would appear to be more appropriate. Since $U(t)$ takes only m distinct values during a complete period, the natural type of approximant to use is then

$$u(t) \simeq U_0 + \sum_{k=1}^{m-1} U_k \sin(k\omega t + \psi_k)$$

with

$$U_k \exp(\mathrm{i}\psi_k) = \frac{\mathrm{i}}{\pi}\int_0^{2\pi} u(t)\exp(-\mathrm{i}k\omega t)\,\mathrm{d}(\omega t)$$

and correspondingly

$$e(t) \simeq r - G(0)N_0U_0 - \sum_{k=1}^{m-1} \mathrm{Im}[G(\mathrm{i}k\omega)N_kU_k\exp\{\mathrm{i}(k\omega t+\psi_k)\}]$$

where N_0 and N_k are components of the MIDF for $f(u)$.

Alternatively, we can apply the describing function method in a different way, by combining the digital part of the system together with the nonlinearity, taking a harmonic approximant for $e(t)$, and calculating the corresponding expression for $f(u(t))$ from Fourier analysis. In this version, the assumption of a simple biased sinusoidal form for the signals may be more legitimate, as the effect of the harmonics generated by the sampling process is, to some extent, implicitly included. We thus take

$$y(t) \simeq Y_0 + Y_1\sin(\omega t + \phi)$$

so that the input to the nonlinearity is approximated by

$$f(u(t)) \simeq G^{-1}(0)Y_0 + \mathrm{Im}[G^{-1}(\mathrm{i}\omega)Y_1\exp\{\mathrm{i}(\omega t+\phi)\}]$$

and the error signal by

$$e(t) \simeq r - Y_0 - Y_1\sin(\omega t+\phi).$$

Now, since the system is supposed to be in a state of steady oscillation, persisting over all time rather than starting at $t = 0$, the relation between u and e is given by

$$u(kT_\mathrm{s}) = \sum_{j=-\infty}^{k} d_{k-j}\,e(jT_\mathrm{s})$$

which is to be substituted into the expression for $f(u)$ in order to calculate the Fourier coefficients. This then gives the equations

$$G^{-1}(0)\,Y_0 = \frac{1}{m}\sum_{k=0}^{m-1} f(u(kT_\mathrm{s}))$$

$$G^{-1}(\mathrm{i}\omega)\,Y_1\exp(\mathrm{i}\phi) = \frac{1-\exp(-2\pi\mathrm{i}/m)}{\pi}\sum_{k=0}^{m-1} f(u(kT_\mathrm{s}))\exp\left(\frac{-2\pi\mathrm{i}k}{m}\right)$$

on taking account of the piecewise-constant nature of $u(t)$. Consequently, we now have a set of equations from which (Y_0, Y_1, ϕ) can be determined, although the connection with describing functions for $f(u)$ is lost in this approach.

Instead of making an approximation, however, we can in fact perform an exact analysis, along the lines of Tsypkin's method as described in Chapter 4. To do this, we express the piecewise-constant signal $u(t)$ in the form

$$u(t) = \sum_{k=0}^{m-1} u(kT_s)\, v(t-kT_s)$$

where $v(t)$ denotes the pulse train shown in Fig. 7.4, which has the Fourier expansion

$$v(t) = \frac{1}{m} + \frac{1}{\pi} \sum_{k=0}^{\infty} \frac{\sin(k\omega t) - \sin\{k(\omega t - 2\pi/m)\}}{k}$$

since $T_s = T/m$. With the Tsypkin functions defined in Chapter 4, it then follows, after much algebra, that

$$y(kT_s) = \sum_{j=0}^{m-1} f(u(jT_s))\, \delta_{j-k}$$

where the coefficients are given by

$$\delta_k = \frac{G(0)}{m} + \frac{1}{2}\,\mathrm{Im}\left\{\Lambda\left(\frac{2\pi k}{m}, \omega\right) - \Lambda\left(\frac{2\pi\,(k+1)}{m}, \omega\right)\right\}$$

and hence we can obtain a set of m equations for the m distinct levels of $u(t)$, using the relation

$$u(kT_s) = \sum_{j=-\infty}^{k} d_{k-j}\,\{r - y(jT_s)\}$$

which arises from the error sampling. Moreover, this procedure can

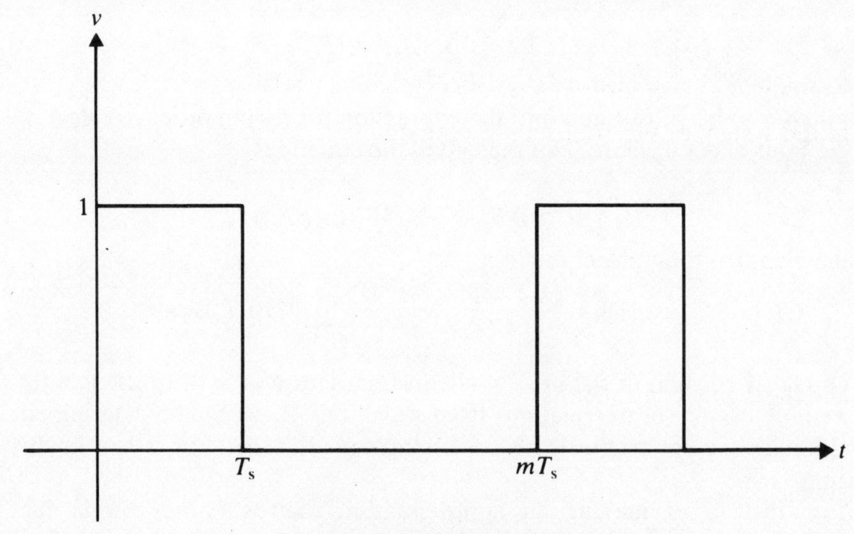

Fig. 7.4 The pulse train $v(t)$.

immediately be generalised to incorporate a nonlinear dependence of e on r and y, caused by a sensor nonlinearity and/or by the 'quantisation' effect of sampling with a finite word-length, as represented in Fig. 7.5.

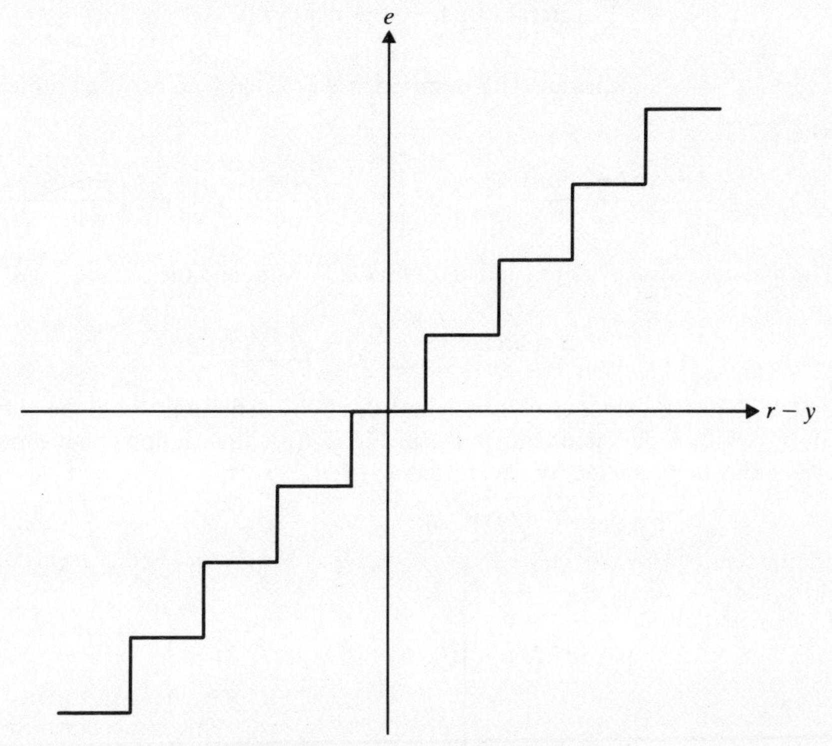

Fig. 7.5 Quantisation nonlinearity.

Example 7.4 Sampled-data system containing a relay

In the system of Fig. 7.2, suppose the nonlinear element is an ideal relay, so that

$$f(u) = \text{sgn}(u)$$

the plant has the transfer function

$$G(s) = \frac{1}{s(s+1)}$$

and the digital controller has

$$D(z) = 1$$

giving

$$u(kT_s) = e(kT_s)$$

for all $k \geq 0$. Taking the DIDF approximation for $y(t)$, with r held fixed and $Y_0 = r$, the conditions for a limit cycle of period mT_s become

$$\sum_{k=0}^{m-1} \operatorname{sgn}\left\{ Y_1 \sin\left(\frac{2\pi k}{m} + \phi\right)\right\} \doteq 0$$

$$i\omega(i\omega+1)\, Y_1 \exp(i\phi) = -F(Y_1,\phi)$$

where

$$F(Y,\phi) = \frac{1-\exp(-2\pi i/m)}{\pi} \sum_{k=0}^{m-1} \operatorname{sgn}\left\{ Y \sin\left(\frac{2\pi k}{m} + \phi\right)\right\} \exp\left(\frac{-2\pi i k}{m}\right).$$

The first condition evidently indicates that m is even, and the second gives

$$\phi = \arctan\left(\frac{mT_s}{2\pi}\right) + \angle F(Y_1,\phi)$$

on taking phases, since $\omega = 2\pi/mT_s$ and $Y_1 > 0$, by definition. From the plot of $\angle F$, which is independent of Y_1, in Fig. 7.6, it then follows that limit cycles may be predicted for several values of m.

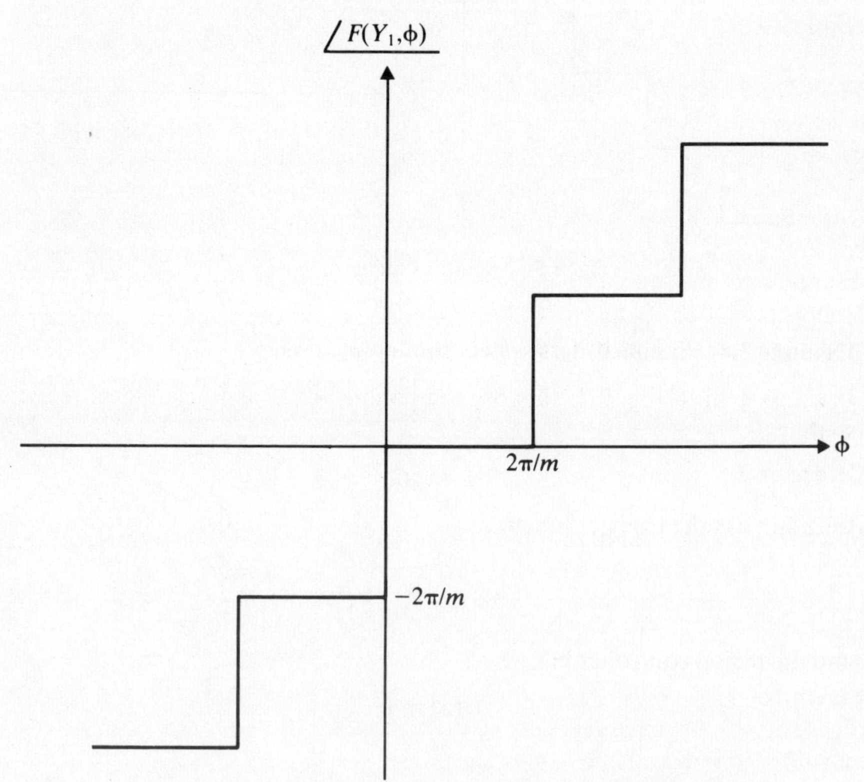

Fig. 7.6 Plot of $\angle F$ against ϕ for even m, in Example 7.4.

APPENDIX 1 NOTES AND REFERENCES

Here we collect some information regarding the sources of the material contained in the above chapters, together with comments and references which may be useful to those wishing to pursue various topics further into the technical literature.

Chapter 1

The contents of this chapter are mainly standard. It will be apparent, however, that in order to appreciate this subject, it is necessary to have a sound grasp of linear system theory, as found in Barnett (1975) or Brockett (1970), and an adequate mathematical background, such as is provided by Rosenbrock & Storey (1970). Also, since the subject is treated from a control engineering viewpoint, the reader should be familiar with the terminology and methods used in this field, as described in D'Azzo & Houpis (1960) or Richards (1979). A good general treatment of nonlinear control theory is given by Gibson (1963).

Chapter 2

Much of the material here (on linearisation, the phase portrait, and limit cycles) is fairly orthodox and could be found in, for example, J. L. Willems (1970), whose nomenclature we use for the different types of singular point, in contrast to some other authors, such as Hirsch & Smale (1974). The more advanced topics of chaotic motion and strange attractors are covered by Mees (1981) and, at a rather abstract level, Smale (1980). With regard to the examples included in the chapter, Example 2.1 is a simple model from power system engineering, as described by Elgerd (1971). The famous oscillator model in Example 2.2 was originally proposed by Van der Pol (1926), and the ecological system in Example 2.3 is based on Lotka (1924) and Volterra (1931), of which more modern developments are discussed by Montroll (1972). Incidentally, the functions denoted by V in Examples 2.1 and 2.3 are special cases of Lyapunov functions, which we discuss at some length in Chapter 5. Example 2.4 originates from a study of hydrodynamic flow by Lorenz (1963), and the same model was later shown by Haken (1975) to describe the occurrence of certain instabilities in the operation of lasers. Example 2.5 is taken from the discussion by Jacobs (1977) of dynamical models which have been used in attempts to explain the observed history of the earth's magnetic field. Finally, the artificial system in Exercise 6, at the end of the chapter, was introduced by Rössler (1977) in a study of chaos.

Chapter 3

The describing function method is covered in many books, particularly detailed treatments being found in Gelb & Vander Velde (1968) and Atherton (1975). Extensions of the method to incorporate multiple harmonics have been made by several authors, notably Mees (1972). There is some confusion of terminology in the literature, since the abbreviation DIDF seems originally to have been used, by West, Douce & Livesley (1956) with regard to a combination of sinusoids at related frequencies, and later restricted to the case of a sinusoid with bias; the phrase two-sinusoidal-input describing function (TSIDF) is also used, when the frequencies are not necessarily commensurable. Applications to the prediction of oscillations in feedback systems containing more than one nonlinearity can be found, for example, in Freeman & Barney (1963) and Gray & Taylor (1979). The construction of error bands in the graphical formulation of the method is discussed in detail by Mees (1984). Examples where the describing function approximation fails, in various ways, have been given by Holtzman (1970), Rapp & Mees (1977), and Mees & Sparrow (1981). The determination of limit cycle stability, for both free and forced systems, is capable of considerably more detailed analysis than we have given it; for example, a treatment based on the so-called incremental describing function (IDF) can be found in Choudhury & Atherton (1974). A study of forced systems, including the topics of jump phenomena, subharmonic oscillations and the suppression of limit cycles, is contained in Lawden (1967). Incidentally, the occurrence of subharmonic solutions, in the large-parameter form of Van der Pol's equation, provided the setting for the earliest studies of chaotic behaviour, by Cartwright & Littlewood (1945) and Levinson (1949).

Chapter 4

Methods for the exact analysis of limit cycles in relay systems were developed in the time domain (point transformation method) by Hamel (1949) and in the frequency domain by Tsypkin (1958); detailed accounts may be found in Gille, Decaulne & Pélegrin (1967) and Atherton (1981), while a very comprehensive treatment is given in Tsypkin (1984). The notation used differs somewhat from one author to another, and does not yet appear to have been standardised; a function equivalent to our $\Lambda(\theta,\omega)$ appears to have been introduced first by Atherton (1966). Example 4.2, in which the point transformation method is used to study chaotic motion in a system with second-order dynamics and a two-valued nonlinearity, is believed to be new, but was inspired by Brockett (1982), where a third-order piecewise-linear system with a memoryless nonlinearity is approximately analysed; this is also closely related to Example 3.7. The stick–slip phenomenon illustrated in Example 4.3 is well known, this particular case being taken from Hagedorn (1978). Variable-structure systems have by now an extensive literature, which can be traced from Itkis (1976) and, especially as regards the connection with adaptive control, Zinober (1981).

Chapter 5

Our account of Lyapunov stability theory follows the same lines as J.L. Willems (1970); a more detailed and extensive treatment is given by Hahn (1967), while Zubov's method, in particular, is developed in Zubov (1957). Applications to power systems are discussed by Rudnick & Brameller (1978), and the use of hyperplanes to approximate the stability boundaries, in this context, is reviewed in Evans (1978) and generalised (using hypersurfaces instead) by Cook & Eskicioglu (1983). The discussion of recurrence properties for central trajectories is based on Lorenz (1963), whose approach derives from Nemytskii & Stepanov (1960). Stability criteria for feedback systems have been presented by numerous authors, including Popov (1962), Sandberg (1964), and Zames (1966), among the earliest; other references can be found in the review by Cook (1979), and in the books of J.C. Willems (1970) and Desoer & Vidyasagar (1975). The concepts and theorems associated with positive-realness are well summarised in the appendices to Landau (1979). Counterexamples to the conjectures of Aizerman (1949) and Kalman (1957) have been found by Fitts (1966) and (in the case of Aizerman) by Dewey & Jury (1965). The relationship between the circle criterion and the describing function was established by Cook (1973). Input–output methods have been widely applied in the derivation of stability conditions, including the off-axis circle criterion by Cho & Narendra (1968), multivariable results in Cook (1975), Kouvaritakis & Husband (1982) and Harris & Valenca (1983), and conditions for the stability of interconnected systems due to Cook (1974); moreover, similar methods were used by Cook (1976) to obtain conditions under which limit cycles cannot occur. There is also another form of input–output representation, applicable to systems with analytic nonlinearities, namely the Volterra series, as described in Rugh (1981), which in principle contains all information about the dynamical behaviour, including stability, but does not seem to reveal it readily in practice.

Chapter 6

The bounds on dynamical variables given here are based on Cook (1980), though those obtained by using impulse-response methods are essentially equivalent to the results of Owens & Chotai (1984); the concept of a conditionally linear system is due to Zakian (1979). Lyapunov functions have been employed, to obtain bounds under persistent disturbances, by Bell (1978), and other approaches which have some relevance to this topic, at least in principle, are those of Hermes (1982), on controllability of nonlinear systems, and Jacobson (1977), concerning the effect of input constraints on linear systems. The use of the describing function concept in dealing with signals of approximately exponential form is discussed in the standard texts in this area, such as Gelb & Vander Velde (1968). Applications of dither are familiar in control engineering, for the effective smoothing of nonlinearities and consequent suppression of unwanted oscillations (signal stabilisation); a rather advanced mathematical treatment of this effect can be found in Zames & Shneydor (1976). Many different forms of adaptive control, including self-tuning and model reference, are covered in Harris & Billings (1981), while an extensive account of the model-reference approach is presented by Landau (1979); the adaptation scheme in Example 6.5 follows the ideas of Parks (1966), and Exercise 4 is taken from Cook & Chen (1982). The literature on the robustness of adaptive systems, which has become an active field of research in recent years, can be traced from Chen & Cook (1984).

Chapter 7

A discussion of the remarkably complicated behaviour, which can be exhibited by simple nonlinear recurrence relations, is to be found in the review by May (1976), with particular regard to the biological context.

Example 7.3 is based on a recursive algorithm described in Wellstead & Zarrop (1983). A general introduction to sampled-data control is provided by Kuo (1963), and methods for the prediction of limit cycles, in nonlinear sampled-data systems, are given in McNamara & Atherton (1984).

References

Aizerman M.A. (1949) 'On a problem concerning the global stability of dynamic systems', *Usp. Mat. Nauk.* **4**, 187–8.

Atherton D.P. (1966) 'Conditions for periodicity in control systems containing several relays', *Proc. 3rd IFAC World Congress*, Pergamon Press, Oxford, Paper 28E.

Atherton D.P. (1975) *Nonlinear Control Engineering,* Van Nostrand Reinhold, Wokingham.

Atherton D.P. (1981) *Stability of Nonlinear Systems,* John Wiley, Chichester.

Barnett S. (1975) *Introduction to Mathematical Control Theory,* Oxford University Press, Oxford.

Bell D.J. (1978) 'Finite time stability of a high energy beam storage ring' *Proc. 7th IFAC World Congress*, Pergamon Press, Oxford, pp. 1711–15.

Brockett R.W. (1970) *Finite Dimensional Linear Systems,* John Wiley, Chichester.

Brockett R.W. (1982) 'On conditions leading to chaos in feedback systems', *Proc. IEEE Conf. on Decision and Control,* IEEE Press, New York, pp. 932–6.

Cartwright M.L. & Littlewood J.E. (1945) 'On nonlinear differential equations of the second order', *J. London Math. Soc.* **20** 180–9.

Chen Z.J. & Cook P.A. (1984) 'Robustness of model-reference adaptive control systems with unmodelled dynamics', *Int. J. Control* **39** 201–14.

Cho Y.S. & Narendra K.S. (1968) 'An off-axis circle criterion for the stability of feedback systems with a monotonic nonlinearity', *IEEE Trans.* **AC–13** 413–16.

Choudhury S.K. & Atherton D.P. (1974) 'Limit cycles in high-order nonlinear systems', *Proc. IEE* **121** 717–24.

Cook P.A. (1973) 'Describing function for a sector nonlinearity', Proc. IEE **120** 143–4.

Cook P.A. (1974) 'On the stability of interconnected systems', *Int. J. Control* **20** 407–15.

Cook P.A. (1975) 'Circle criteria for stability in Hilbert space', *SIAM J. Control* **13** 593–610.

Cook P.A. (1976) 'Conditions for the absence of limit cycles' *IEEE Trans.* **AC–21** 339–45.

Cook P.A. (1979) 'Circle theorems and functional analytic methods in stability theory', *Proc. IEE* **126** 616–22.

Cook P.A. (1980) 'On the behaviour of dynamical systems subject to bounded disturbances,' *Int. J. Syst. Sci.* **11** 159–70.

Cook P.A. & Chen Z.J. (1982) 'Robustness properties of model-reference adaptive control systems', *IEE Proc.* Pt.D **129** 305-309.

Cook P.A. & Eskicioglu A.M. (1983) 'Transient stability analysis of electric power systems by the method of trangent hypersurfaces', *IEE Proc.* PtC **130** 183–93.

D'Azzo J.J. & Houpis C.H. (1960) *Control System Analysis and Synthesis*, McGraw-Hill, New York.

Desoer C.A. & Vidyasagar M. (1975) *Feedback Systems: Input–Output Properties*, Academic Press, Florida.

Dewey A.G. & Jury E.I. (1965) 'A note on Aizerman's conjecture', *IEEE Trans.* **AC–10** 482–3.

Elgerd O.I. (1971) *Electric Energy Systems Theory*, McGraw-Hill, New York.

Evans F.J. (1978) 'Prospects for dynamic security monitoring in large scale electric power systems', *Proc. 7th IFAC World Congress,* Pergamon Press, Oxford pp 1–14.

Fitts R.E. (1966) 'Two counterexamples to Aizerman's conjecture', *IEEE Trans.* **AC–11** 553–6.

Freeman E.A. & Barney G.C. (1963) 'Divergent oscillations and their excitation in control systems with two saturation-type non-linear elements', *Proc IEE* **110** 1096–1106.

Gelb A. & Vander Velde W.E. (1968) *Multiple-Input Describing Functions and Nonlinear System Design*, McGraw-Hill, New York.

Gibson J.E. (1963) *Nonlinear Automatic Control*, McGraw-Hill, New York.

Gille J.C., Decauline P. & Pélegrin M. (1967) *Méthodes d'Etude des Systèmes Asservis non Linéaires*, Dunod, Paris.

Gray J.O. & Taylor P.M. (1979) 'Computer-aided design of multivariable non-linear control systems using frequency domain techniques' *Automatica* **15** 281–97.

Hagedorn P. (1978) *Nichtlineare Schwingungen*, Akademische Verlagsgesellschaft, Engl. Transl. (1982) *Non-linear Oscillations* Oxford University Press, Oxford.

Hahn W. (1967) *Stability of Motion*, Springer, Berlin.

Haken H. (1975) 'Analogy between higher instabilities in fluids and lasers', *Phys. Lett.* **53A** 77–8.

Hamel B. (1949) 'Contribution à l'étude mathematique des systèmes de réglage par tout-ou-rien', *C.E.M.V.17*, Service technique aeronautique.

Harris C.J. & Billings S.A., Eds. (1981) *Self-tuning and Adaptive Control: Theory and Application*, Peter Peregrinus, Stevenage.

Harris C.J. & Valenca J.M.E. (1983) *The Stability of Input–Output Dynamical Systems*, Academic Press, Florida.

Hermes H. (1982) 'Control of systems which generate decomposable Lie algebras', *J. Differ. Eqns.* **44** 166–87.

Hirsch M.W. & Smale S. (1974) *Differential Equations, Dynamical Systems and Linear Algebra*, Academic Press, Florida.

Holtzman J.M. (1970) *Nonlinear System Theory*, Prentice-Hall, Englewood Cliffs.

Itkis U. (1976) *Variable Structure Systems*, John Wiley, Chichester.

Jacobs J.A. (1977) 'The earth's core and geomagnetism', *Bull. IMA* **13** 86–90.
Jacobson D.H. (1977) *Extensions of Linear-Quadratic Control, Optimization and Matrix Theory*, Academic Press, Florida.

Kalman R.E. (1957) 'Physical and mathematical mechanisms of instability in nonlinear automatic control systems', *Trans. ASME J. Basic Engng* **79** 553–66.
Kouvaritakis B. & Husband R.K. (1982) 'Multivariable circle criteria: an approach based on sector considerations' *Int. J. Control* **35** 227–54.
Kuo B.C. (1963) *Analysis and Synthesis of Sampled-Data Control Systems*, Prentice-Hall, Englewood Cliffs.

Landau Y.D. (1979) *Adaptive Control: the Model Reference Approach*, Marcell Dekker, New York, Basel.
Lawden D.F. (1967) *Mathematics of Engineering Systems*, Methuen, London.
Levinson N. (1949) 'A second-order differential equation with singular solutions', *Ann. Math.* **50** 127–53.
Lorenz E.N. (1963) 'Deterministic nonperiodic flow', *J. Atmos. Sci* **20** 130–41.
Lotka A.J. (1924) *Elements of Physical Biology*, reprinted as (1956) *Elements of Mathematical Biology*, Dover, New York.

May R.M. (1976) 'Simple mathematical models with very complicated dynamics' *Nature* **261** 459–67.
McNamara O.P. & Atherton D.P. (1984) 'Limit cycles in nonlinear sampled-data systems' *Proc. 9th IFAC World Congress,* Pergamon Press, Oxford pp. 214–19.
Mees A.I. (1972) 'The describing function matrix', *J. Inst. Math. Appl.* **10** 49–67.
Mees A.I. (1981) *Dynamics of Feedback Systems*, John Wiley, Chichester.
Mees A.I. (1984) 'Describing functions: ten years on', *IMAJ. Appl. Math.* **32** 221–33.
Mees A.I. & Sparrow C.T. (1981) 'Chaos', *IEE Proc.* PtD **128** 201–5.
Montroll E.W. (1972) 'On coupled rate equations with quadratic nonlinearities', *Proc. Natl Acad. Sci. USA* **69** 2532–6.

Nemytskii V.V. & Stepanov V.V. (1960) *Qualitative Theory of Differential Equations*, Princeton University Press, Princeton.

Owens D.H. & Chotai A. (1984) 'Stability and performance deterioration due to modelling errors' Billings S.A. *et al.* (Eds) in *Nonlinear System Design,* Peter Peregrinus, Stevenage.

Parks P.C. (1966) 'Liapunov redesign of model reference adaptive control systems', *IEEE Trans.* **AC–11** 362–7.
Popov V.M. (1962) 'Absolute stability of nonlinear systems of automatic control', *Autom. & Rem. Control* **22** 857–75.

Rapp P.E. & Mees A.I. (1977) 'Spurious prediction of limit cycles in a nonlinear feedback system using the describing function method', *Int. J. Control* **26** 821–9.

Richards R.J. (1979) *An Introduction to Dynamics and Control*, Longman, Harlow.

Rosenbrock H.H. & Storey C. (1970) *Mathematics of Dynamical Systems*, Nelson, Walton-on-Thames.

Rössler O.E. (1977) 'Continuous chaos' in Haken H. (Ed.) *Synergetics: a Workshop*, Springer, Berlin.

Rudnick H. & Brameller A. (1978) 'Transient security assessment methods', *Proc. IEE* **125** 135–40.

Rugh W.J. (1981) *Nonlinear System Theory: the Volterra–Wiener Approach*, John Hopkins University Press, Baltimore.

Sandberg I.W. (1964) 'On the L_2-boundedness of solutions of nonlinear functional equations', *Bell Syst. Tech. J.* **43** 1581–9.

Smale S. (1980) *The Mathematics of Time*, Springer, Berlin.

Tsypkin Y.Z. (1958) *Theorie der Relais Systeme der Automatischen Regelung*, Oldenbourg.

Tsypkin Y.Z. (1984) *Relay Control Systems*, Cambridge University Press, Cambridge.

Van der Pol B. (1926) 'On relaxation oscillations', *Phil. Mag.* **2** 978–92.

Volterra V. (1931) *'Leçon sur la Théorie Mathematique de la Lutte pour la Vie*, Gauthier Villars, Paris.

Wellstead P.E. & Zarrop M.B. (1983) 'Self-tuning regulators: non-parametric algorithms', *Int. J. Control* **37** 787–807.

West J.C., Douce J.L. & Livesley R.K. (1956) 'The dual input describing function and its use in the analysis of nonlinear feedback systems', *Proc. IEE* Pt B **103**, 463–74.

Willems J.C. (1970) *The Analysis of Feedback Systems*, MIT Press, Massachussetts.

Willems J.L. (1970) *Stability Theory of Dynamical Systems*, Nelson, Walton-on-Thames.

Zakian V. (1979) 'New formulation for the method of inequalities', *Proc. IEE* **126** 579–84.

Zames G. (1966) 'On the input–output stability of nonlinear time-varying feedback systems', *IEEE Trans.* **AC–11** 228–39, 465–76.

Zames G. & Shneydor N.A. (1976) 'Dither in nonlinear systems', *IEEE Trans.* **AC–21** 660–7.

Zinober A.S.I. (1981) 'Controller design using the theory of variable structure systems', in Harris C.J. & Billings S.A. (Eds) *Self-Tuning and Adaptive Control* Peter Peregrinus, Stevenage, pp. 206–29.

Zubov V.I. (1957) *The Methods of Lyapunov and their Applications*, Leningrad University Press, Leningrad.

APPENDIX 2 SOLUTIONS TO EXERCISES

Chapter 2

1 With state variables $x_1 = \delta$ and $x_2 = \dot{\delta}$, the state equations become

$$\dot{x}_1 = x_2$$

$$\dot{x}_2 = \frac{P_m - P_e \sin x_1 - Kx_2}{H}$$

so that the equilibrium points are the same as in Example 2.1. Linearisation around $(\hat{x}_1, 0)$ gives the coefficient matrices

$$\mathbf{A} = \begin{bmatrix} 0 & 1 \\ \dfrac{-P_e \cos \hat{x}_1}{H} & \dfrac{-K}{H} \end{bmatrix}$$

$$\mathbf{B} = \begin{bmatrix} 0 \\ \dfrac{1}{H} \end{bmatrix} \qquad \mathbf{C} = [1 \quad 0],$$

with the input ΔP_m and output $\Delta\delta$. Therefore the transfer function is

$$G(s) = \frac{1}{Hs^2 + Ks + P_e \cos \hat{x}_1}$$

where $\sin \hat{x}_1 = P_m/P_e$. Hence

$$G(s) = \frac{1}{Hs^2 + Ks \pm \sqrt{(P_e^2 - P_m^2)}}$$

where the upper and lower signs correspond to the stable and unstable equilibria, respectively.

2 A suitable region is illustrated in Fig. A.1. The inner boundary is a circle with centre (0,0) and radius <1, while the outer boundary is composed of a set of straight line segments and circular arcs. Considering only the upper half, because of the symmetry, it is clear that trajectories cross the segment AB to the right, CD downwards, and the arcs BC and DE in the inward direction. Also, on the segment EF, we have

$$\frac{dx_2}{dx_1} = \epsilon(1-x_2^2) - \frac{x_1}{x_2} \leqslant \epsilon - \frac{x_1}{x_2} \leqslant \epsilon - b$$

while the slope of EF itself is $-1/(a-b)$. Now, since C lies on the isocline of slope zero, we obtain

$$a = b\sqrt{\{1 + \epsilon^2 b^2 (1+b^2)\}}$$

and so, by taking b suficiently large, we can make

$$b-\epsilon > \frac{1}{a-b}$$

so that trajectories crossing EF always do so in the downward direction. All the conditons required for the application of the Poincaré–Bendixson theorem are then satisfied.

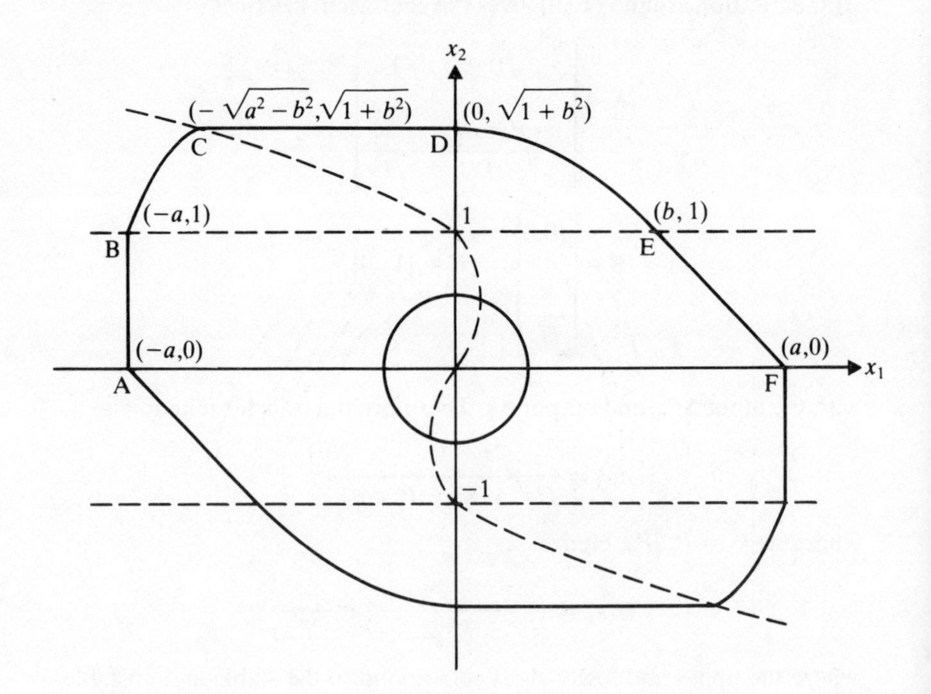

Fig. A.1

3 After substituting $\dot{x}_1$ into the expression for $\dot{x}_2$, the state equations become

$$\begin{aligned}\dot{x}_1 &= x_1 - x_1x_2 - \mu x_1^2\\ \dot{x}_2 &= (1-\tau)x_1x_2 + \tau x_1x_2^2 + \tau\mu x_1^2x_2 - x_2\end{aligned}$$

Thus the equilibrium points are:

$$(0,0), \qquad \left(\frac{1}{\mu}, 0\right), \quad (1,1-\mu).$$

Linearisation around $(1,1-\mu)$ gives $\Delta\dot{x} = \mathbf{A}\Delta x$, with

$$\mathbf{A} = \begin{pmatrix} -\mu & -1 \\ (1+\tau\mu)(1-\mu) & \tau(1-\mu) \end{pmatrix}$$

Therefore, $\det(s\mathbf{I}-\mathbf{A}) = s^2 + (\mu+\tau\mu-\tau)s + (1-\mu)$.

Hence, for this equilibrium to be stable, we require

$$\mu \leqslant 1, \qquad \tau \leqslant \frac{\mu}{1-\mu}.$$

If the first of these conditions is satisfied but the second is not, then trajectories near the equilibrium point will diverge from it, although others, starting from distant points, approach it, leading to the formation of a limit cycle, as in Fig. A.2.

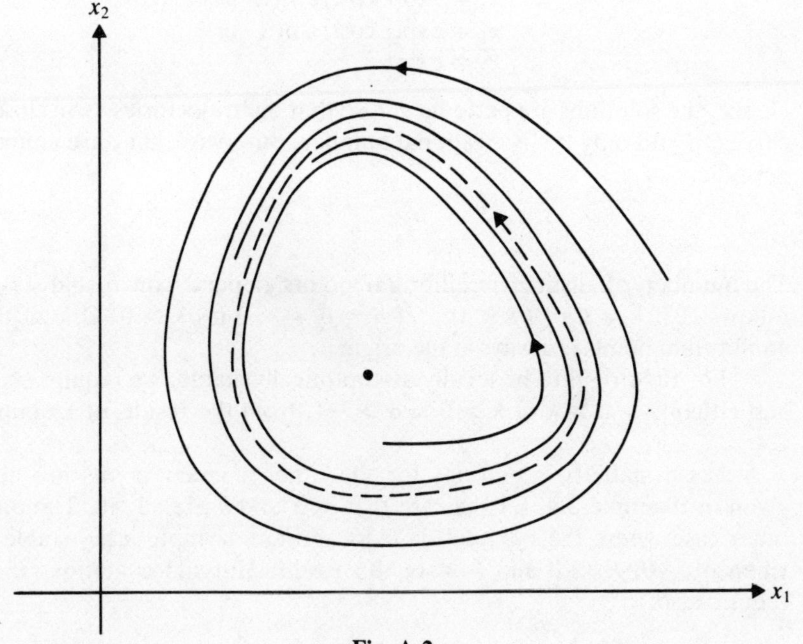

Fig. A.2

4 Defining

$$y = \sqrt{(x_1^2 + x_2^2)}$$

we have

$$\dot{y} = -x_3 y$$
$$\dot{x}_3 = \ln y$$

so that

$$(\ln y)^2 + x_3^2 = c^2$$

where c is a constant. For any fixed value of c, this equation describes a torus, of noncircular cross-section, on which the trajectories lie. Also, defining

$$\theta = \arctan (x_2/x_1)$$

we get

$$\dot{\theta} = -\upsilon$$

whence

$$\theta = \alpha - \upsilon t$$

for some constant α. Therefore, with a suitable choice of time origin, the solution for $x(t)$ can be written

$$x_1 = \exp(c \cos t) \cos(\alpha - \upsilon t)$$
$$x_2 = \exp(c \cos t) \sin(\alpha - \upsilon t)$$
$$x_3 = c \sin t.$$

Hence, the solutions are periodic in t, so that the trajectories form closed curves, if and only if υ is a rational number; otherwise, they are almost-periodic.

5 The number of distinct equilibrium points depends on b and λ, as follows: 1 if $\lambda = 0$ or $b\lambda < 0$; 2 if $b = 0 \neq \lambda$; 3 if $b\lambda > 0$. One of the equilibrium points is always at the origin.

For the origin to be locally asymptotically stable, we require $b > 0$ and either $\sigma > 0 > \lambda$ or $\lambda > 0 > \sigma > -1$, from the results of Example 2.4.

Local stability conditions for the other singular points are also given in Example 2.4, for the case that σ, λ and b are all >0. The only other case where these equilibria exist and are asymptotically stable is when $\sigma > 0$, $\lambda < 0$ and $b < 0$; the Routh–Hurwitz conditions then require also

$$-\lambda > \sigma + 1 > -b$$

and

$$\lambda(b+1-\sigma) + (\sigma+1)(\sigma+1+b) < 0$$

for asymptotic stability.

It is thus not possible for all three equilibria to be asymptotically stable, for the same set of parameter values.

6 This system always has an equilibrium point at the origin. Linearisation around this point gives the characteristic equation

$$(s+a)\{s^2+(c-1)s + (b-c)\} = 0$$

and so the local asymptotic stability conditions are

$$a > 0, \quad b > c > 1.$$

Also, if

$$d = a\left(1 - \frac{b}{c}\right) > 0$$

there are two more singular points, at

$$\left(\pm\sqrt{d}, 1 - \frac{b}{c}, \pm\frac{b\sqrt{d}}{c}\right).$$

From the symmetry of the system, we need only linearise about the first of these points, giving a plant matrix

$$\mathbf{A} = \begin{bmatrix} \frac{b}{c} & -\sqrt{d} & -1 \\ 2\sqrt{d} & -a & 0 \\ b & 0 & -c \end{bmatrix}$$

so that

$$\det(s\mathbf{I}-\mathbf{A}) = s^3 + \left(a+c-\frac{b}{c}\right)s^2+a\left(c+2-\frac{3b}{c}\right)s + 2a(c-b)$$

whence, using $d > 0$, the Routh–Hurwitz conditions for local asymptotic stability become

$$c > 0, \quad a > \frac{b}{c} - c,$$

$$a\left\{\left(a+c - \frac{b}{c}\right)\left(c+2 - \frac{3b}{c}\right) - 2(c-b)\right\} > 0.$$

7 Expressing ω_1 and ω_2 in terms of x_3 and α as in Example 2.5, with α now being a dynamical variable, we have

$$\dot{\alpha} = \frac{q_1 - q_2}{2} - \epsilon\alpha$$

so that

$$\alpha \rightarrow \alpha_0 = \frac{q_1 - q_2}{2\epsilon}$$

as $t \rightarrow \infty$. In analysing the equilibria, we can thus assume that this limiting value has been attained, so that the state equations become

$$\begin{aligned}
\dot{x}_1 &= -\mu_1 x_1 + (x_3 + \alpha_0) x_2 \\
\dot{x}_2 &= -\mu_2 x_2 + (x_3 - \alpha_0) x_1 \\
\dot{x}_3 &= q - x_1 x_2 - \epsilon x_3
\end{aligned}$$

where $q = (q_1 + q_2)/2$. There is then always a singular point at $(0,0,q/\epsilon)$, and the characteristic equation of the corresponding linearised model is

$$(s+\epsilon)\{(s+\mu_1)(s+\mu_2) - q_1 q_2/\epsilon^2\} = 0$$

on taking the values of q and α_0 into account.

Also, if $|q| > \epsilon\gamma$, where

$$\gamma = \sqrt{(\mu_1\mu_2 + \alpha_0^2)}$$

there are two more equilibrium points. Taking q to be positive, for definiteness, these are located at $(\beta_1, \beta_2, \gamma)$ and $(-\beta_1, -\beta_2, \gamma)$, where

$$\beta_1 = \sqrt{\frac{(\gamma+\alpha_0)(q-\epsilon\gamma)}{\mu_1}}$$

$$\beta_2 = \sqrt{\frac{(\gamma-\alpha_0)(q-\epsilon\gamma)}{\mu_2}}$$

with similar expressions in the case that q is negative. Linearising around the first of these points, which are evidently related by symmetry, we obtain the plant matrix

$$\mathbf{A} = \begin{pmatrix} -\mu_1 & \gamma+\alpha_0 & \beta_2 \\ \gamma-\alpha_0 & -\mu_2 & \beta_1 \\ -\beta_2 & -\beta_1 & -\epsilon \end{pmatrix}$$

whence the characteristic equation $\det(s\mathbf{I} - \mathbf{A}) = 0$ becomes

$$s(s+\epsilon)(s+\mu_1+\mu_2) + (q-\epsilon\gamma)\left\{\left(\frac{\gamma+\alpha_0}{\mu_1} + \frac{\gamma-\alpha_0}{\mu_2}\right)s + 4\gamma\right\} = 0.$$

Since the condition $|q| > \epsilon\gamma$ is equivalent to

$$q_1 q_2 > \epsilon^2 \mu_1 \mu_2$$

it follows that the equilibria at $(\beta_1, \beta_2, \gamma)$ and $(-\beta_1, -\beta_2, \gamma)$ exist only when the one at $(0,0,q/\epsilon)$ is unstable. This can only happen if q_1 and q_2 have the same sign, and ϵ is sufficiently small.

Chapter 3

1 With input

$$u = U\sin(\omega t)$$

the output is given by

$$y = \begin{cases} u-a, & -\alpha \leqslant \omega t \leqslant \pi/2, \\ U-a, & \pi/2 \leqslant \omega t \leqslant \pi-\alpha, \\ u+a, & \pi-\alpha, \leqslant \omega t \leqslant 3\pi/2, \\ -U+a, & 3\pi/2 \leqslant \omega t \leqslant 2\pi-\alpha, \end{cases}$$

where

$$\alpha = \arcsin\left(1-\frac{2a}{U}\right).$$

Therefore, by using symmetry properties, the SIDF can be written in the form

$$\begin{aligned} \mathrm{Re}N &= \frac{2}{\pi U}\left(\int_{-\alpha}^{\pi/2}(U\sin\theta-a)\sin\theta\mathrm{d}\theta+(U-a)\int_{\pi/2}^{\pi-\alpha}\sin\theta\mathrm{d}\theta\right) \\ &= \frac{2}{\pi}\left\{\frac{\pi}{4}+\frac{\alpha}{2}-\frac{\sin(2\alpha)}{4}+\left(1-\frac{2a}{U}\right)\cos\alpha\right\} \\ &= \frac{1}{2}+\frac{\alpha}{\pi}+\frac{\sin(2\alpha)}{2\pi} \end{aligned}$$

$$\mathrm{Im}N = \frac{-4a(U-a)}{\pi U^2}$$

which agrees with the general expression for the imaginary part of a describing function, due to a hysteresis loop. Thus, substituting for α gives

$$N = \frac{1}{2}+\frac{1}{\pi}\left[\arcsin\left(1-\frac{2a}{U}\right)+2\left(1-\frac{2a}{U}\right)\sqrt{\left\{\frac{a}{U}\left(1-\frac{a}{U}\right)\right\}}\right] - \frac{4\mathrm{i}}{\pi}\frac{a}{U}\left(1-\frac{a}{\mathrm{U}}\right).$$

2 The SIDF for the dead zone nonlinearity is

$$N = 1 - \frac{2}{\pi}\left\{ \arcsin\left(\frac{a}{U}\right) + \frac{a}{U}\sqrt{\left(1 - \frac{a^2}{U^2}\right)}\right\}$$

for $U \geqslant a$, with $N = 0$ for $U < a$. Hence, the graph of $-1/N$, in the complex plane, goes along the negative real axis from $-\infty$ to -1, as U goes from 0 to ∞. Also, the Nyquist plot of $G(s)$, as illustrated in Fig. A.3, crosses the real axis when $\operatorname{Im} G(i\omega) = 0$, so that $\omega = 1$ and the crossing point is at $-K/2$.

Thus, if

$$K > 2$$

the loci intersect, and so a limit cycle is predicted. The value of the constant reference signal is irrelevant, since $G(s)$ contains an integrator, which justifies the use of the SIDF rather than the DIDF. However, the locus of $-1/N$ crosses the Nyquist plot from left to right, seen from the direction of increasing ω, so the limit cycle is expected to be unstable.

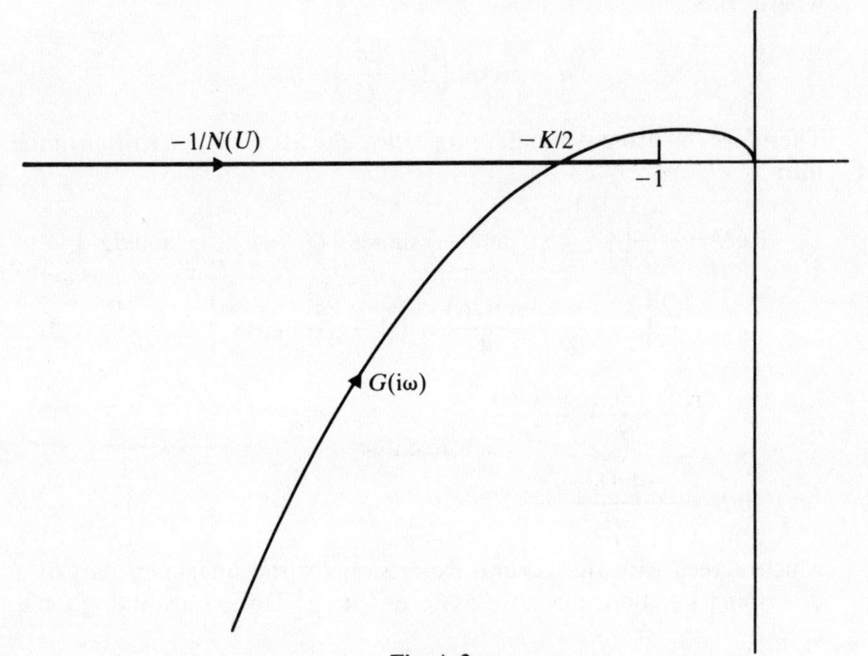

Fig. A.3

3 Since the nonlinearity is single-valued, the equation for the limit cycle frequency, in the DIDF approximation, is

$$\operatorname{Im} G(i\omega) = 0$$

which gives

$$\omega^2 = 1 + \frac{2\zeta}{\lambda}$$

and hence

$$\operatorname{Re} G(i\omega) = -\lambda/2\zeta.$$

The other conditions for a limit cycle then become

$$(1 + N_0)U_0 = r$$

$$1 - \frac{\lambda}{2\zeta} N_1 = 0$$

using the fact that $G(0) = 1$, while the DIDF components are given by

$$N_0 = U_0 + \frac{U_1^2}{2U_0}, \qquad N_1 = 2U_0.$$

Solving these equations, we obtain

$$U_0 = \frac{\zeta}{\lambda}, \qquad U_1 = \sqrt{\{2(r - U_0 - U_0^2)\}},$$

so that the solution exists only if

$$r > \frac{\zeta}{\lambda}\left(1 + \frac{\zeta}{\lambda}\right).$$

Also, since U_0 is a decreasing function of U_1, so is N_1, and the graph of the effective negative inverse describing function thus crosses the Nyquist plot from right to left, giving a prediction of a stable limit cycle.

4 Taking the reference input as

$$r = R\sin(\omega t)$$

and approximating the loop signal u by

$$u \simeq U\sin(\omega t + \psi)$$

we obtain

$$\{1 + G(i\omega)\, N\}\, U\exp(i\psi) = R$$

with

$$G(i\omega) = \frac{-1}{\omega^2} - \frac{i}{\omega}$$

and

$$N = \frac{3U^2}{4}.$$

Hence, eliminating ψ, we get

$$R^2 = U^2 \left\{ \left(1 - \frac{3U^2}{4\omega^2}\right)^2 + \frac{9U^4}{16\omega^2} \right\}.$$

Now, setting the derivative of the right-hand side (with respect to U^2) equal to zero gives

$$\frac{27}{16}\left(\frac{1+\omega^2}{\omega^4}\right) U^4 - \frac{3U^2}{\omega^2} + 1 = 0$$

which has real distinct positive roots for U^2 if

$$0 < \omega < \frac{1}{\sqrt{3}}.$$

Consequently, for this range of ω, there exists a range of values of R such that the equation for U^2 has three distinct roots, two of which are predicted to correspond to stable conditions, between which jumps can take place, while the third is associated with an unstable mode of operation.

Chapter 4

1 A state-space representation for the linear element is given by

$$\begin{aligned}\dot{x}_1 &= Kf(u)\\ \dot{x}_2 &= -x_2 + Kf(u)\\ y &= x_1 - x_2\end{aligned}$$

where

$$u = r - y$$

and $f(u)$ denotes the relay characteristic of Fig. 3.9. Since r is constant, we can absorb it into the definition of x_1, which is equivalent to setting $r = 0$. The switching lines in the state plane are then $x_2 = x_1 \pm a$, as shown in Fig. A.4.

Thus, suppose a trajectory starts at the point $(p, p-a)$, where the relay output is just switching to $f(u) = -1$. Solving the state equations, we find

$$\begin{aligned}x_1 &= p - Kt\\ x_2 &= (p-a+K)\exp(-t)-K\end{aligned}$$

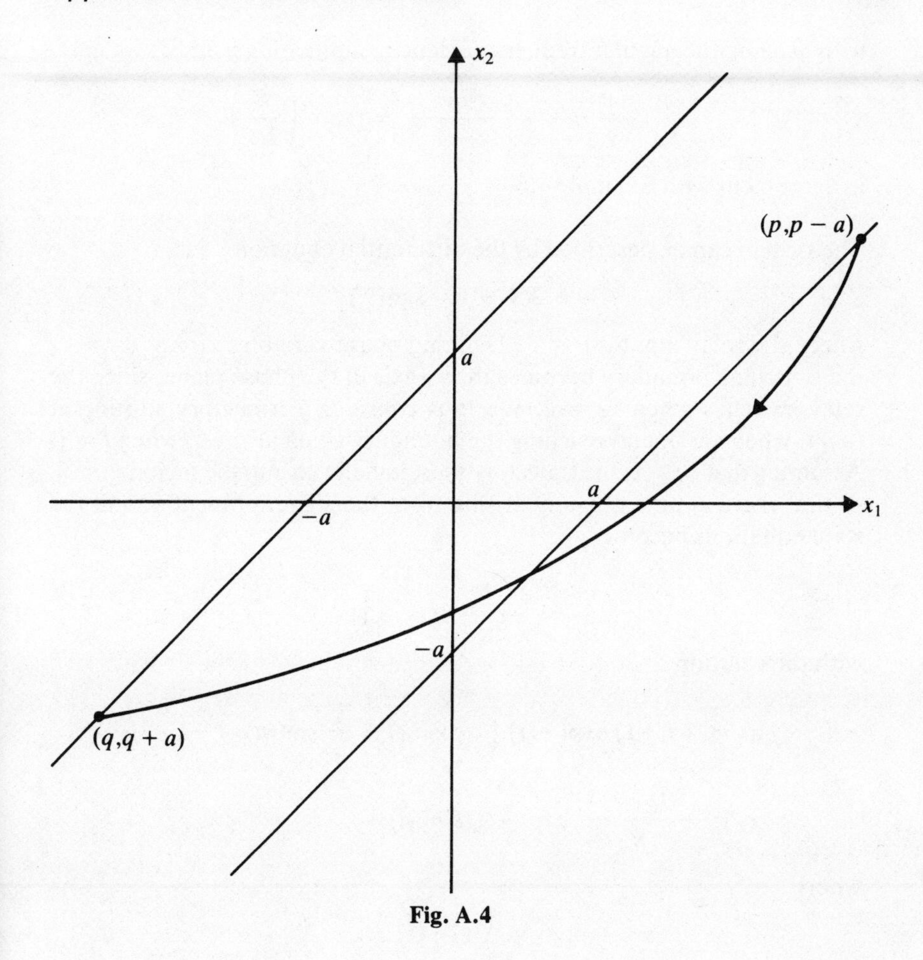

Fig. A.4

so that, if the trajectory reaches the other switching line at $(q,q+a)$ after time τ, we have

$$q+a+K = (p-a+K)\exp(-\tau)$$

where

$$\tau = \frac{p-q}{K}.$$

Now, the condition for a limit cycle is

$$q = -p$$

and τ is then the half-period, whence

$$\tau = \frac{\pi}{\omega} = \frac{2p}{K}$$

with ω being the angular frequency. Hence, eliminating p and q, we get

$$\frac{\pi}{2\omega} - \frac{a}{K} = \frac{\exp(\tau)-1}{\exp(\tau)+1} = \tanh\left(\frac{\pi}{2\omega}\right)$$

in agreement with Example 4.6.

2 The system can be described by the differential equation

$$\ddot{v} + 2\zeta\dot{v} + v = \mathrm{sgn}(\dot{v})$$

where the relay input is $u = \dot{v}$. Defining phase variables $x_1 = v$, $x_2 = \dot{v}$, the switching boundary becomes the x_1-axis in the phase plane, since the relay switches when $x_2 = 0$. We thus consider a trajectory starting at $(p_1,0)$ when $t = 0$, and reaching the boundary again at $(p_2,0)$ when $t = \tau$. Assuming that $p_1 > 1$, the trajectory must initially go into the region $\dot{v} < 0$, so that there is no ambiguity arising from the signum function, and the state equations become

$$\dot{x}_1 = x_2$$
$$\dot{x}_2 = -x_1 - 2\zeta x_2 - 1$$

with the solution

$$x_1 = (p_1+1)\exp(-\zeta t)\left(\cos(\gamma t) + \frac{\zeta}{\gamma}\sin(\gamma t)\right) - 1$$

$$x_2 = \frac{-(p_1+1)}{\gamma}\exp(-\zeta t)\sin(\gamma t)$$

where

$$\gamma = \sqrt{(1-\zeta^2)}.$$

Consequently, the next switching time is $\tau = \pi/\gamma$, giving

$$p_2 = -(p_1+1)\exp\left(-\frac{\zeta\pi}{\gamma}\right) - 1$$

as plotted in the Lemeré diagram of Fig. A.5, from which we see that there is a stable limit cycle with

$$p_1 = -p_2 = \coth\left(\frac{\zeta\pi}{2\gamma}\right) > 1$$

and period $2\tau = 2\pi/\gamma$, as found in Example 4.7.

On the other hand, the describing function method predicts a limit cycle with angular frequency $\omega = 1$, at which $G(i\omega)$ crosses the real axis, and corresponding period 2π, so the prediction is accurate only in the limit $\zeta \to 0$ and fails completely for $\zeta > 1$.

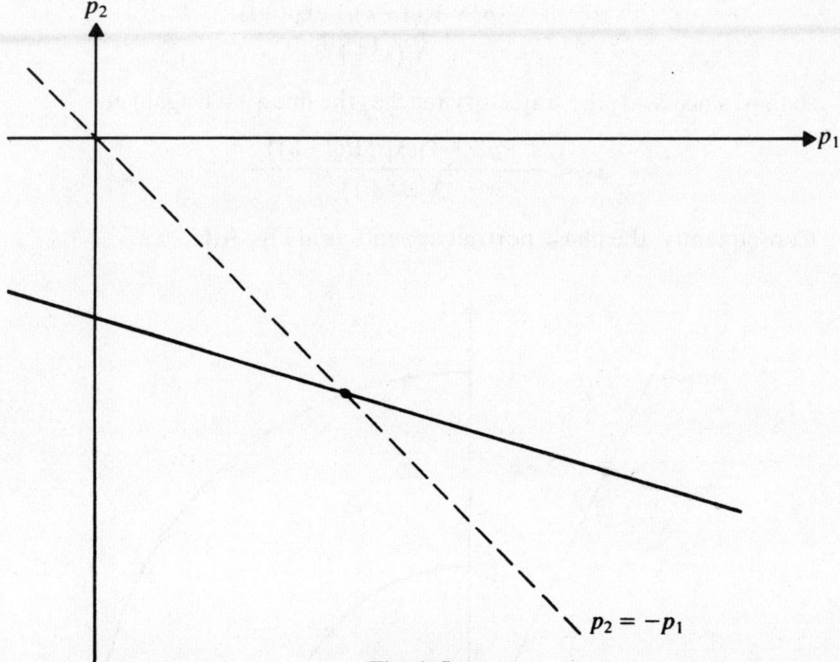

Fig. A.5

3 For the region

$$\sigma x_1 > 0$$

the state equations give

$$\frac{\mathrm{d}}{\mathrm{d}t}\left(x_1^2 + x_2^2 \right) = 0$$

so the trajectories are arcs of circles, centred at the origin. Hence, a trajectory starting at $(0,p)$, with $p > 0$, reaches the switching line $\sigma = 0$ at $(p/\sqrt{(\lambda^2+1)}, -p\lambda/\sqrt{(\lambda^2+1)})$. Then, in the region

$$\sigma x_1 < 0$$

we have

$$\dot{x}_1 = x_2$$

$$\dot{x}_2 = -x_1 - 2x_2$$

whose solution, for the given initial conditions, is

$$x_1 = \frac{p\{1 + (1-\lambda)t\} \exp(-t)}{\sqrt{(\lambda^2+1)}}$$

$$x_2 = \frac{-p\{\lambda + (1-\lambda)t\}\exp(-t)}{\sqrt{(\lambda^2+1)}}$$

so that, since $\lambda>1$, the trajectory reaches the line $x_1 = 0$ again at

$$x_2 = \frac{-p(\lambda-1)\exp\{1/(1-\lambda)\}}{\sqrt{(\lambda^2+1)}}.$$

Consequently, the phase portrait appears as in Fig. A.6.

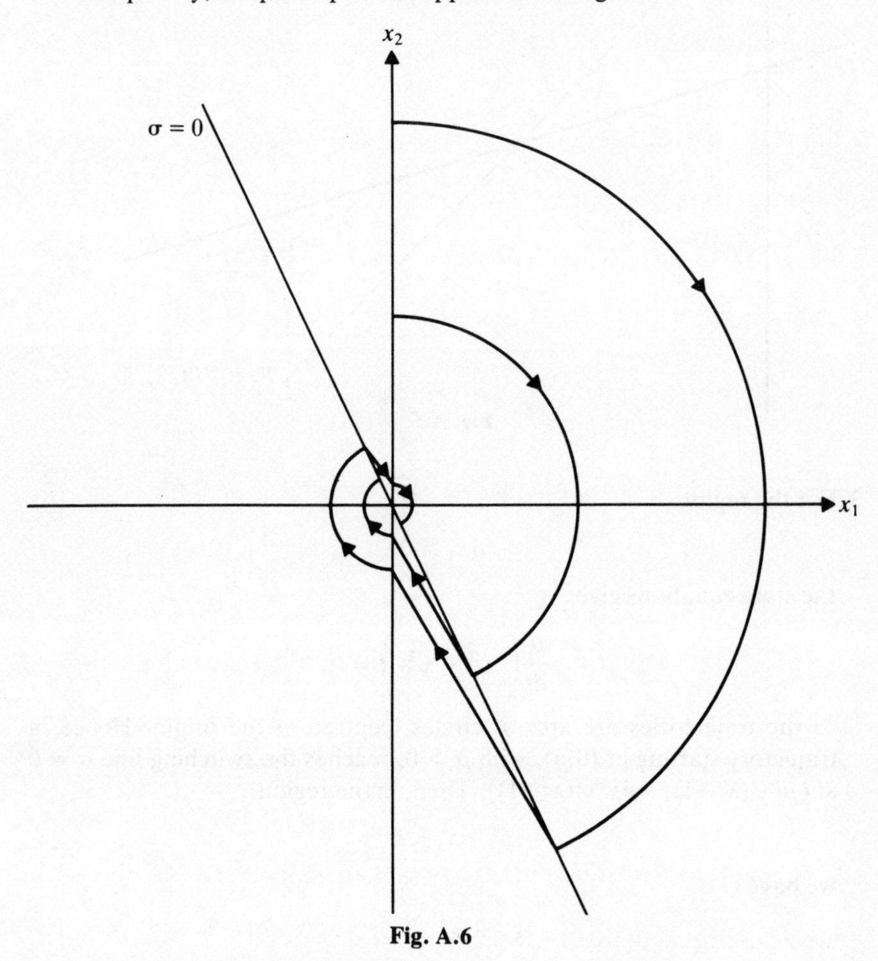

Fig. A.6

4 Since the relay characteristic is given by Fig. 4.18 with $a = b$, the limit cycle conditions are

$$\text{Im}\{\Lambda_0(0,\omega) - \Lambda_0(\theta,\omega)\} = -\frac{2b}{h}$$

$$\mathrm{Im}\{\Lambda_0(0,\omega) - \Lambda_0(-\theta,\omega)\} = \frac{2b}{h}$$

$$\mathrm{Re}\{\Lambda_0(0,\omega) - \Lambda_0(\theta,\omega)\} \leqslant \frac{-K}{\omega}$$

$$\mathrm{Re}\{\Lambda_0(0,\omega) - \Lambda_0(-\theta,\omega)\} \leqslant \frac{-\mathrm{K}}{\omega}$$

using the fact that

$$\rho = -K.$$

From the partial fraction form of $G(s)$ in Example 4.1, the required Tsypkin functions are given by

$$\mathrm{Re}\Lambda_0(\theta,\omega) = \frac{-\mathrm{sgn}(\theta)}{\omega} - \frac{(1+K)}{\omega}\exp\left(\frac{\theta}{\omega}\right)\left\{\tanh\left(\frac{\pi}{2\omega}\right) - \mathrm{sgn}(\theta)\right\}$$

$$\mathrm{Im}\Lambda_0(\theta,\omega) = \frac{1}{\omega}\left(|\theta| - \frac{\pi}{2}\right) + (1+K)\left\{1 - \exp\left(\frac{\theta}{\omega}\right)\right\}\mathrm{sgn}(\theta)$$

$$+(1+K)\exp\left(\frac{\theta}{\omega}\right)\tanh\left(\frac{\pi}{2\omega}\right)$$

so that the limit cycle equations become

$$(1+K)\left\{1 - \exp\left(\frac{\theta}{\omega}\right)\right\}\left\{\tanh\left(\frac{\pi}{2\omega}\right) - \mathrm{sgn}(\theta)\right\} - \frac{|\theta|}{\omega} = \frac{-2b}{h}$$

$$(1+K)\left\{1 - \exp\left(\frac{-\theta}{\omega}\right)\right\}\left\{\tanh\left(\frac{\pi}{2\omega}\right) + \mathrm{sgn}(\theta)\right\} - \frac{|\theta|}{\omega} = \frac{2b}{h}$$

which can be rewritten as

$$(1+K)\{\exp(\tau)-1\}\left\{1-\tanh\left(\frac{T}{4}\right)\right\} = \tau - \frac{2b}{h}$$

$$(1+K)\{1-\exp(-\tau)\}\left\{1+\tanh\left(\frac{T}{4}\right)\right\} = \tau + \frac{2b}{h}$$

taking into account that

$$\theta = \omega\tau > 0$$

and $\omega = 2\pi/T$. Eliminating T, we then get

$$1+K = \frac{b}{h} + \frac{\tau}{2}\coth\left(\frac{\tau}{2}\right)$$

which agrees with the results of Example 4.1, while we also obtain

$$(1+K)\tanh\left(\frac{T}{4}\right)=\frac{\tau}{2}+\frac{b}{h}\coth\left(\frac{\tau}{2}\right).$$

Now, the inequality conditions reduce to

$$\{\exp(\tau)-1\}\left\{\tanh\left(\frac{T}{4}\right)-1\right\}\leqslant 0$$

$$\frac{K-1}{1+K}\leqslant\{1-\exp(-\tau)\}\tanh\left(\frac{T}{4}\right)-\exp(-\tau)$$

which, using the above expressions for K and T, are equivalent to

$$\frac{b}{h}\leqslant\frac{\tau}{2}$$

$$\exp(\tau)-1\geqslant\tau$$

the second of which is automatically satisfied, while the first requires

$$K\geqslant\frac{b}{h}\left\{1+\coth\left(\frac{b}{h}\right)\right\}-1$$

as used in Example 5.5.

5 With

$$G(s)=\frac{1}{s+1}$$

the appropriate Tsypkin conditions are

$$\text{Im}\Lambda_0(0,\omega)=-a$$

$$\text{Re}\Lambda_0(0,\omega)\leqslant\frac{1}{\omega}$$

where

$$\text{Re}\Lambda_0(0,\omega)=\frac{1}{\omega}\tanh\left(\frac{\pi}{2\omega}\right)$$

$$\text{Im}\Lambda_0(0,\omega)=-\tanh\left(\frac{\pi}{2\omega}\right)$$

so the equality condition is satisfied when

$$\tanh\left(\frac{\pi}{2\omega}\right)=a.$$

This clearly requires

$$a < 1$$

and the inequality constraint then also holds.

However, the Nyquist plot of $G(s)$ lies entirely in the right half-plane, and thus cannot intersect the graph of $-1/N$ for the relay element, which is shown in Fig. 3.15. Hence, the describing function method fails to predict the limit cycle at all.

6 Setting $\phi = \theta + \omega$, to allow for the delay factor, we can write the relevant Tsypkin functions as

$$\mathrm{Re}\Lambda_0(\theta,\omega) = \frac{-\mathrm{sgn}(\phi)}{\omega}$$

$$\mathrm{Im}\Lambda_0(\theta,\omega) = \frac{1}{\omega}\left(|\phi| - \frac{\pi}{2}\right)$$

for $|\phi| < \pi$, with

$$\Lambda_0(\pi+\theta,\omega) = -\Lambda_0(\theta,\omega).$$

Consequently, the limit cycle equation

$$\mathrm{Im}\Lambda_0(0,\omega) = 0$$

gives

$$\omega = (k + \tfrac{1}{2})\pi$$

for any non-negative integer k. The corresponding half-period, $1/(k+1/2)$, cannot be an integral submultiple of the delay, which is 1, so the time-derivative of the relay output is continuous at the switching points, and hence the inequality condition becomes

$$\mathrm{Re}\Lambda_0(0,\omega) \leqslant 0$$

so that k must be even, giving periods $\{4, \tfrac{4}{5}, \tfrac{4}{9}, \ldots\}$.

Chapter 5

1 The equilibrium points $x = \hat{x}$ are given by

$$\hat{x}^3 - \hat{x} = r$$

which has three distinct real roots if $|r| < 2/3\sqrt{3}$, but only one if $|r| > 2/3\sqrt{3}$. Also, setting $\tilde{x} = x - \hat{x}$ and linearising, we get

$$\dot{\tilde{x}} = (1 - 3\hat{x}^2)\tilde{x}$$

so that the equilibrium at $\hat{x}$ is asymptotically stable for $\hat{x}^2 > 1/3$, and unstable for $\hat{x}^2 < 1/3$. Thus, when

$$|r| > \frac{2}{3\sqrt{3}}$$

the sole equilibrium point is globally asymptotically stable, that is to say, its maximal domain of attraction is the entire state space, which, for this system, is just the real line. However, when

$$|r| < \frac{2}{3\sqrt{3}}$$

there are three equilibria, of which the middle one is unstable while the other two are (locally) asymptotically stable. Consequently, the state space is then divided, at the unstable equilibrium, into two semi-infinite line segments, each of which is the maximal domain of attraction for one of the stable equilibrium points.

2 Using the Lyapunov function

$$V = y^2 - \tfrac{2}{3} y^3 + \dot{y}^2$$

we have

$$\dot{V} = -4\zeta\dot{y}^2 \leqslant 0$$

and $\dot{V}$ does not vanish identically along any non-degenerate trajectory, so the interior of any closed region $\{V \leqslant \text{constant}\}$ is a domain of attraction. The contours of V are clearly given by

$$\dot{y}^2 = \tfrac{2}{3} y^3 - y^2 + c$$

where c is constant, and the largest closed one is obtained with $c = 1/3$, so that

$$\dot{y}^2 = \frac{(y-1)^2(2y+1)}{3}.$$

Expressed in terms of the state variables used in Example 5.2, this becomes

$$x_2 = \frac{\zeta x_1}{\gamma} \pm \frac{x_1 - 1}{\gamma}\sqrt{\left(\frac{2x_1+1}{3}\right)}$$

which is shown in Fig. A.7, for the same value of ζ as in Fig. 5.5, by way of comparison.

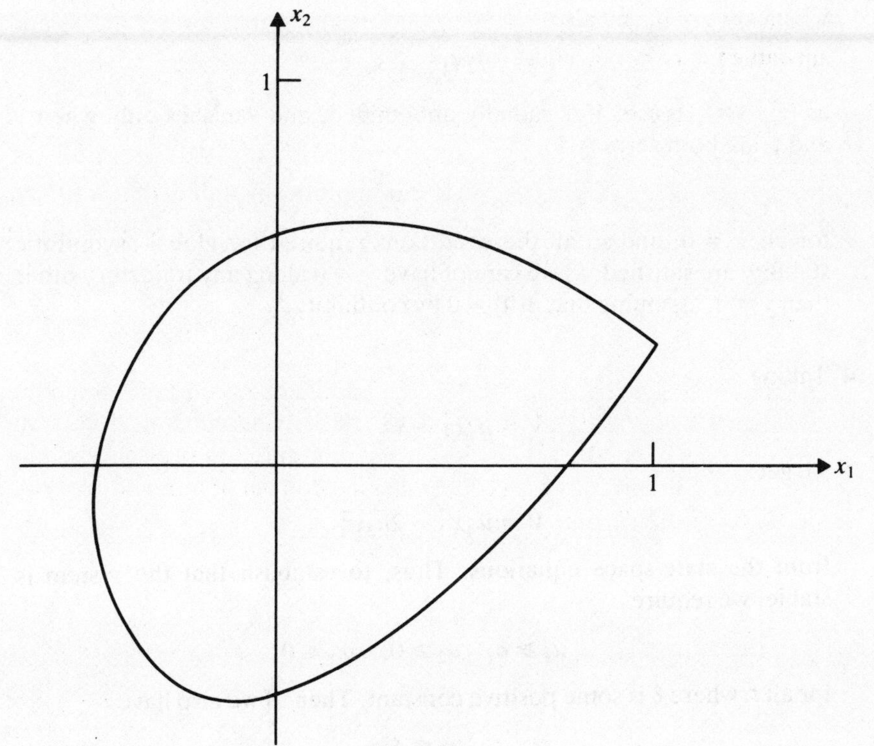

Fig. A.7

3 Defining

$$F(y) = \int_0^y f(v)\,\mathrm{d}v$$

we take the Lyapunov function

$$V = \frac{\dot{y}^2}{2} + F(y)$$

so that

$$\dot{V} = -\dot{y}\,h(\dot{y})$$

from the differential equation governing the system. Now, since

$$\frac{f(y)}{y} \geqslant \epsilon$$

for all $y \neq 0$, with ϵ being a positive constant, it follows that

$$F(y) > 0$$

whenever $y \neq 0$, and also

$$F(y) \to \infty$$

as $|y| \to \infty$. Hence, V is radially unbounded, and vanishes only when y and $\dot{y}$ are both zero. Also,

$$\dot{V} < 0$$

for all $\dot{y} \neq 0$, and so all the conditions required for global asymptotic stability are satisfied, as we cannot have $\dot{y} \equiv 0$ along any trajectory other than $y \equiv 0$, assuming that $h(0) = 0$ by continuity.

4 Taking

$$V = u_1 x_1^2 + x_2^2$$

we get

$$\dot{V} = \dot{u}_1 x_1^2 - 2u_2 x_2^2$$

from the state-space equations. Thus, to establish that the system is stable, we require

$$u_1 \geqslant \delta, \quad u_2 \geqslant 0, \quad \dot{u}_1 \leqslant 0$$

for all t, where δ is some positive constant. Then, if we also have

$$u_1 \leqslant K$$

where K is constant, for all t, V is decrescent and the system is uniformly stable.

We note that it is not possible to prove asymptotic stability, with this Lyapunov function, since we should need

$$\dot{u}_1 \leqslant -\epsilon$$

for all t, with some constant $\epsilon > 0$, so that u_1 would eventually become negative.

5 From the transfer function

$$G(s) = \frac{1}{s(s+1)}$$

we obtain

$$\text{Re}\,G(i\omega) = \frac{-1}{1+\omega^2}$$

$$\text{Im}\,G(i\omega) = \frac{-1}{1+\omega^2}$$

so that the modified polar plot consists of a straight line segment joining the point $(-1,-1)$ to the origin, as shown in Fig. A.8. Hence, for any constant $\beta > 0$, the modified polar plot lies entirely to the right of any straight line of positive slope $\leqslant 1$, passing through $(-1/\beta, 0)$. Consequently, if the nonlinear element in the feedback system satisfies

$$0 < yf(y) < \beta y^2$$

for all $y \neq 0$, where β is an arbitrary positive constant, then the system is globally asymptotically stable, by the Popov criterion.

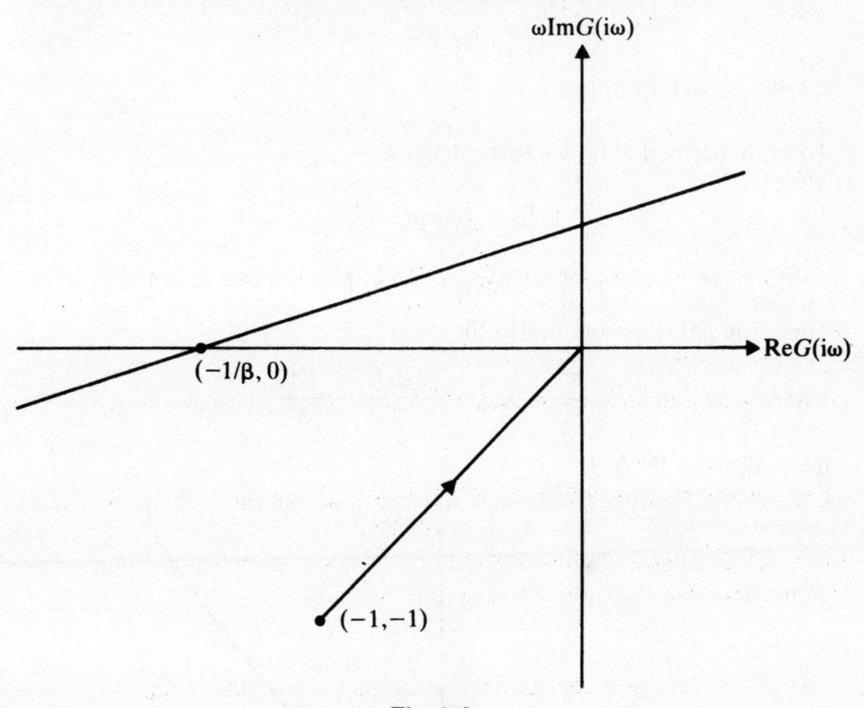

Fig. A.8

Chapter 6

1 Replacing the linear element transfer function in Example 6.1 by

$$G(s) = \frac{2s+1}{s^2}$$

we find that the closed-loop transfer function of the linearised system is

$$G_K(s) = \frac{2s+1}{(s+1)^2}$$

for $K = 1$, with the corresponding impulse-response function

$$g_K(t) = (2-t)\exp(-t)$$

so that

$$\gamma(\infty) = \int_0^2 (2-t)\exp(-t)\mathrm{d}t + \int_2^\infty (t-2)\exp(-t)\mathrm{d}t$$

$$= 1 + 2\exp(-2) \simeq 1.27$$

The error signal bound is then given by

$$|e(t) - \hat{e}(t)| \leqslant \gamma(\infty)$$

for all t, as in Example 6.1.

2 To evaluate the EIDF, we first calculate

$$U = \int_0^\infty [\{E_0 + E\exp(-\theta)\}^3 - E_0^3]\mathrm{d}\theta$$

$$= 3E_0^2E + 3/2\, E_0E^2 + 1/3\, E^3$$

Then, the EIDF is obtained as the ratio

$$\frac{U}{E} = 3E_0^2 + \frac{3E_0E}{2} + \frac{E^2}{3}$$

illustrated in Fig. A.9.

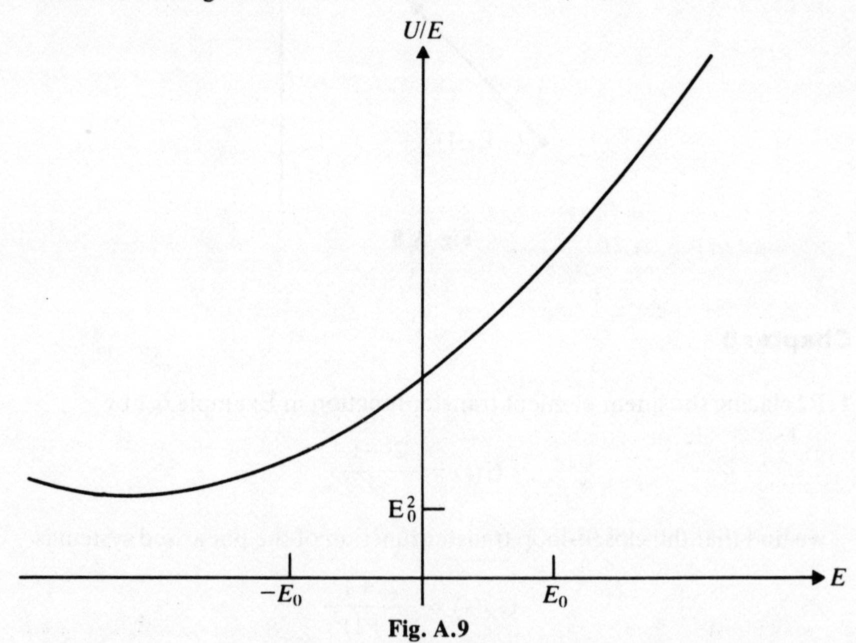

Fig. A.9

3 With $|e| \geqslant W$, where W is the dither amplitude, $f(e)$ is linear throughout the range swept by the dither signal, so the effective nonlinearity is just $f(e)$ itself. We thus take $|e| < W$ in the following analysis.

(i) For a square-wave dither, we have

$$\bar{f}(e) = \frac{f(e+W) + f(e-W)}{2} = \frac{e+W}{2}.$$

(ii) For a sinusoidal dither, we define

$$\alpha = \arcsin(e/W)$$

so that

$$\begin{aligned}\bar{f}(e) &= \frac{1}{2\pi}\int_{-\alpha}^{\pi+\alpha} (e + W\sin\theta)\mathrm{d}\theta \\ &= \left(\frac{1}{2} + \frac{\alpha}{\pi}\right)e + \frac{W\cos\alpha}{\pi} \\ &= \left\{\frac{1}{2} + \frac{1}{\pi}\arcsin\left(\frac{e}{W}\right)\right\}e + \frac{\surd(W^2-e^2)}{\pi}.\end{aligned}$$

(iii) For a triangular-wave dither, we get

$$\begin{aligned}f(e) &= \frac{1}{2W}\int_{-e}^{W} (e+w)\mathrm{d}w \\ &= \frac{1}{2W}\left((W+e)e + \frac{W^2-e^2}{2}\right) \\ &= \frac{(W+e)^2}{4W}.\end{aligned}$$

The graphs of the effective nonlinearities in these three cases are plotted in Fig. A.10.

4 Under the specified control law, the operation of the plant is described by the equation

$$\dot{y} = aK(r-y)$$

while the output y_R of the reference model satisfies

$$T\dot{y}_R = r - y_R.$$

Hence, defining an error signal

$$e_0 = y - y_R$$

we obtain

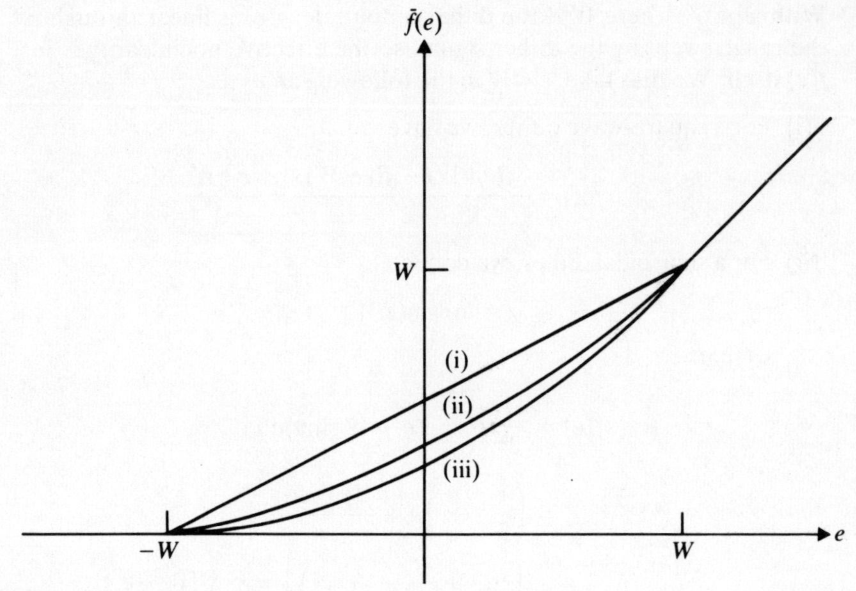

Fig. A.10

$$\dot{e}_0 = -\frac{\dot{e}_0}{T} + \left(aK - \frac{1}{T}\right)(r-y)$$

$$= -\frac{e_0}{T} + a\phi(r-y)$$

where

$$\phi = K - \frac{1}{aT}.$$

Thus, for a Lyapunov function of the form

$$V = e_0^2 + \lambda\phi^2$$

where λ is a positive constant, it follows that

$$\dot{V} = -\frac{2}{T}e_0^2 + 2\phi\{\lambda\dot{\phi} + ae_0(r-y)\}$$

and so the adaptation law

$$\dot{\phi} = \mu(y-r)e_0$$

gives

$$\dot{V} = -\frac{2}{T}e_0^2 \leqslant 0$$

if we take

$$\lambda = \frac{a}{\mu}.$$

Consequently, we can choose any adaptation law of the above form, provided that $a\mu > 0$.

SUBJECT INDEX